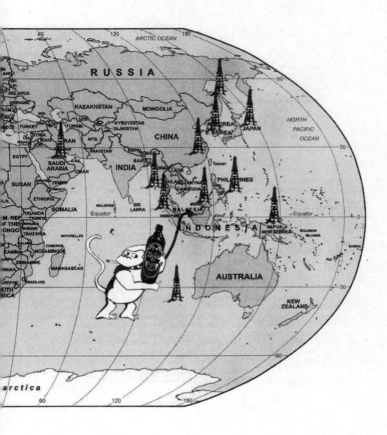

PAUL CARTER **DON'T TELL MUM i WORK ON THE RIGS** SHE THINKS I'M A PIANO PLAYER IN A WHOREHOUSE

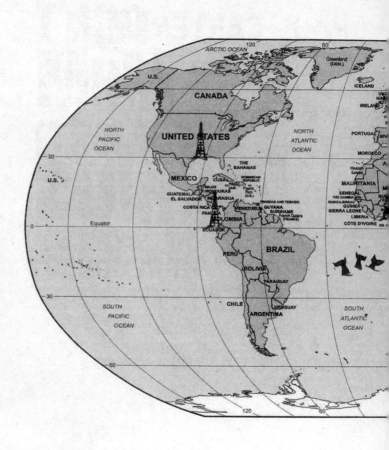

PRAISE FOR
DON'T TELL MUM I WORK ON THE RIGS

"Carter's romper-stomper tour of the world's oil rigs –
a highly enjoyable tale."
The Glasgow Herald

"Great two-fisted writing from the far side of hell."
John Birmingham, bestselling author of
He Died with a Felafel in his Hand and *Dopeland*

"Here's one book you probably can judge by its cover. Paul
Carter's memoir of his life as an oil man in some of the
freakiest, most lawless locations in the world is not for the faint-
hearted, as the name – borrowed from an old bumper sticker
and now possibly my favourite book title ever – suggests. But if
you've got the stomach for exploding monkeys, explosive
dysentery, gunfights, hijacks and brothels staffed entirely by
dwarves, you're in for a treat. *Don't Tell Mum I Work on the Rigs*
takes the reader on a white-knuckle ride around the oilfields of
Nigeria, Russia, Asia, the Middle East and South America,
barely stopping for breath as it scrambles from one audacious
adventure to the next, skipping from near death experience to
side-splitting hilarity so fast you hope he's kept a few anecdotes
up his sleeve for the next book."
The Press and Journal, Aberdeen

"Carter's tales are always entertaining and offer a few
unblinking aperçus about Big Oil seen from the inside."
Scotland on Sunday

"A unique look at a gritty game. Relentlessly funny and
obsessively readable."
Phillip Noyce, director of *The Quiet American* and
Clear and Present Danger

PAUL CARTER DON'T TELL MUM I WORK ON THE RIGS

SHE THINKS I'M A PIANO PLAYER IN A WHOREHOUSE

nb

NICHOLAS BREALEY
PUBLISHING

LONDON · BOSTON

This edition first published in the UK by
Nicholas Brealey Publishing in 2007
Reprinted 2007 (three times), 2008 (twice)

First published 2006

3–5 Spafield Street
Clerkenwell, London
EC1R 4QB, UK
Tel: +44 (0)20 7239 0360
Fax: +44 (0)20 7239 0370

20 Park Plaza, Suite 1115A
Boston
MA 02116, USA
Tel: (888) BREALEY
Fax: (617) 523 3708

www.nicholasbrealey.com
www.paulcarter.net.au

ISBN: 978-1-85788-377-0

British Library Cataloguing in Publication Data
A catalogue record for this book is available from the
British Library.

Printed in the UK by Clays Ltd, St Ives plc.

CONTENTS

ACKNOWLEDGEMENTS

First and foremost, I'd like to thank Erwin Herczeg, for watching my back on more occasions than I care to mention and proving to be the perfect role model. Thanks also to Drew Gardenier for getting me started in the first place, and letting me get away with a damn sight more than I deserved to. Special thanks to Sally and Simon Dominguez and Lou and Doug Frost, and Susan Coghill, without whom this book would never have happened. My thanks to all the boys who backed me up, covered my arse and listened to my bullshit over the years; you know who you are. To the team at Allen & Unwin, especially Jo Paul, Lou Johnson, Alexandra Nahlous and Catherine Milne—thank you and sorry I can't spell. Last, but certainly not least, all my love to Clare, Elinor, Johannes, Allan and France. God bless.

CALENTURE

A name formerly given to a tropical fever or delirium suffered by sailors after long periods at sea, who imagine the ocean to be green fields and desire to leap into them.

LAND RIG

JACK-UP RIG

TENSION LEG PLATFORM

DRILL SHIP

DRILLING & PRODUCTION FIXED PLATFORM

SEMI-SUBMERSIBLE RIG

There are many different types of drilling rigs with multiple variations on their capabilities and adaptability to the environment in which they are drilling in. I have not gone into the technical aspect of this industry. But so you have a basic overview this page gives a rough idea of the different rigs that are out there and their most obvious components.

JACK-UP RIG

1. DRILL FLOOR
2. DERRICK
3. PIPE RACK
4. ACCOMMODATION BLOCK, OFFICES, RADIO ROOM, GALLEY, SHOWERS, RECREATION ROOM
5. HELI-DECK

PROLOGUE

THE MAN SITTING NEXT to me looked panicked; I tried to hold on, beads of nervous sweat forming on my forehead. My knuckles went from white to blue, I bit into my upper lip. About halfway through the take-off climb, the inevitable happened: my backside let go. I yelped in complete horror as two piping-hot spurts of poo shot down my trouser legs . . . I'd just lost my arse on a commercial airliner . . . oh my God.

I fumbled at the seatbelt, my IV bag dangling from my mouth. The man next to me ran away as I hobbled down the fuselage towards the nirvana of a business-class toilet. Kicking off a metabolic chain reaction, within minutes I had the rest of my crew frantically scurrying towards toilets in immediate and imminent danger of crapping all over the place. And all because some guy in logistics didn't check the bottled water on the rig. What were we thinking, that eight grown men with dysentery could clench all the way from Port Moresby to Singapore? But it was either that or we take the 'death by local hospital' option.

I locked the toilet door, catching my reflection in the mirror as I pulled down my pants and surveyed the horror. I looked so bad for a second I thought I was in there with a pale sweaty stranger. It was one of those unbelievably embarrassing moments in life when you just wish you could go back in time and be ten years old and life was just an endless romp in the park. How did this happen? How in hell did I let this happen?

It was a long flight . . . There's some kid out there in the front row who's going to be disturbed for a while, and that's just because of the smell. I wasn't coming out of that toilet. It gave me time to ponder just how I got to this auspicious point in my ridiculous career path to rapid bowel movement oblivion.

In point of fact it all started when I was about ten but there was no park to romp in . . .

1

ACHTUNG SPITFIRE

1969–87

I WAS BORN IN THE UK to a German mother, an English father, an older sister and a cat called Brim. Brim was an overt snob who would only drink his milk after I had popped any bubbles that floated on the surface. My father would inevitably end up walking by these goings-on and step on the edge of Brim's saucer, sending milk directly up his trouser leg.

My early life was not happy; I don't recall any memories of Dad that make me smile, just overwhelming fear. Brim and I would regularly jostle over the best hiding places, while my father, with his milk-stained trousers, would look for us.

My father was in the Royal Air Force, a navigator. He was a 'children are seen and not heard' kind of dad, and so my sister and I lived a disciplined life. In all our family photos we look like we are having our picture taken for a police line-up.

These should have been the times when life was just an endless romp in the sun and tomorrow didn't matter, when parents were neither a fear nor a worry but something so dependable you would look for that peace of mind in adult life and marry it.

For the Christmas holidays, I was sent to visit my respective grandparents, one year with the Germans, the next with the English. Every year the standard holiday war movie would play on or around Boxing Day, and every year it was something like *The Great Escape* or *The Dirty Dozen*. Throughout the movie, my English grandfather would cheer and clap and when it was over he would pull out his medals and tell me war stories. My German grandfather, on the other hand, would curse and cringe at every standard scene of a single American soldier gunning down countless Germans without once reloading his weapon. When the movie was over, he would crack open a bottle of Schnapps and get blind drunk.

Needless to say we never got together to have nice family dinners with my grandparents.

I think I was around six when I saw *The Magnificent Seven*, the first movie I remember seeing. It had a profound effect on me. My father was away a lot so I took all my male cues from TV. After that movie I wanted to be a cowboy. Just like Steve McQueen.

Then one day, after I had been a cowboy for some months, my father returned from service in Canada with a real Western gun holster. The belt went from just under my nipples to the top of my bellybutton and the holster itself was almost as long as my right leg. All my cap guns just dropped straight through it so I replaced them with my drink cup. It was one of those kiddies' plastic cups with a screw-down lid, it had Charlie Brown on it, but when it was wedged down into the holster all you could see was gun leather and a blue cup handle.

I would parade around the street quick-drawing and slurping cordial. My mother called me 'The Milkshake Sheriff' and wherever she took me the holster came along, Church, Sunday School, the local pool.

The Milkshake Sheriff made only one real enemy in town, a huge Old English sheepdog named 'Benny'. He would spot the Sheriff strutting across the park, bound up, and with one paw knock him down and start shagging him. I hated that dog.

A few years later on a visit to my German grand-parents I sat down to watch the post-Christmas movie and this time it was *The Great Escape*. There was Steve McQueen again. I loved it. By the end of the movie my grandfather was hammered on Schnapps as the Germans lost again and I was ready to trade the holster for a motorbike.

I tooled around the neighbourhood on my pushbike, trying to jump it over people's back fences. I would try to appear surly and indifferent. Looking through one of my mother's magazines one day, I found an article on Steve McQueen. It said he was a man who liked fast bikes and fast women. So I tried to find fast women, but at the age of ten I misunderstood what that meant. I started with the babysitter . . . 'So, you like Fisher Price music baby?' She sent me straight to bed with a firm 'There'll be no Starsky & Hutch for you Mister'.

My mother was a saint; she made up for my father's insanely strict parenting routine with boundless love and affection. One day she got the strength to leave. My sister and I were bundled into the back of the Mini and that was that. Dad ended up leaving the Service and spent the rest of his working life as a directional driller on the rigs; coincidentally my mother ended up working for a major oil company and that led to our eventual emigration to Australia. One of the strange things about the drilling industry is that it is global but very small, and every now and again I run into some old drilling hand who knew my father.

Mum moved to Aberdeen, Scotland, and my sister and I went to a new school; home life, although devoid of the luxuries of my father's house, improved a great deal. We didn't have any money for the kinds of things a ten-year-old boy wants, but we loved each other and Mum made me the man of the house, or rather man of the tiny rundown council flat. I could get away with murder.

When I was fourteen I suddenly started getting bullied

at school. His name was Athel, he was thick and had been held back a year because of it. Athel didn't like my glasses. Mum could not afford to take me to an optometrist so I had National-Health-black-one-size-fits-all nasty grown-up glasses; I looked like a midget Michael Caine.

After Athel was finished I had to tape them together.

My mother's solution to Athel was that I should get a big stick. Instead I talked her into getting me an air rifle. (I had been mildly obsessed with guns since I was five, when I was given a pair of Star Trek pyjamas that came with a phazer gun water pistol.)

The idea was simple. Athel had a gang of boys who kept jumping me and beating me up. He cut the tyres on my bike with a flick knife, and told me he was going to cut my pecker off. So I was going to kill him with my new rifle. It had worked in the war movies . . .

I stole wooden pallets from the loading bay of a nearby supermarket and constructed a hide-out over-looking Athel's backyard. I lay there for hours with my BSA .22 air rifle in hand. I knew Athel stole his father's cigarettes and hid near the shed to smoke them. All I had to do was wait. And sure enough out he came, so I let him have it. The pellet hit him square in the forehead, sending him backwards into the shed door. His screams soon had his father over him, but Athel's sniper-in-the-tree-line story and bleeding head quickly became unimportant as his father realised he had been smoking. Then it was Athel's father who was screaming. I looked on in mute fascination. Athel was not dead, and that's a good thing; he never bothered me again.

Life, as they say, comes down to a few moments. That was one of them. My glasses were never a problem again. A few years after shooting Athel I joined the Gordon Highlanders 2nd Battalion ACF. Thanks to my father I was already totally indoctrinated into the military system—he would check that my toys and clothes were always stowed where they should be, and my room was freakishly tidy—so the military was no surprise. The drill sergeant was scary to the other lads, but compared to my dad he was just a man who yelled a lot. I knew he couldn't hit me. The Highlanders was a great experience, and I managed to fulfil all my childhood firearms needs.

Not long after the Athel incident my mother came home and said she had met a really nice man through her new job with the oil company. His name was John and she wanted to bring him home so my sister and I could meet him. I was happy if she was happy, and my sister was about to move in with her boyfriend so she was happy. John turned out to be great. He was totally relaxed and also treated me like a grown-up, so I didn't give him any shit and he was happy too.

With home life steady at last, I started spending more time hanging around Mum's office or at the workshop talking to the offshore guys when they came in on a crew change. Mostly Americans at that time, they would congratulate me on my polite manners and shove everything from Buck knives to Zippo lighters in my pocket. John would get back from a rig and tell me stories about the strange places he worked in, and always brought me back something cool.

Mum's boss was a larger-than-life character, Jessie Thomas Jackson—JT. A big man in his fifties, impeccably dressed, he, unlike many adults, always shook my hand and despite my mother's protests let me hang out in his giant office. He was remarkably good to me. He would let me drink Coke from his little fridge. I would sit on his massive leather couch, awe-inspired, gazing at the array of rig memorabilia on the shelves.

Sometimes he would look at me over his glasses the way grown-ups do and say, 'C'mon over here Pauli.'

I would run up to his big wooden desk.

'How'd you like to do some work for me?'

'Yessir.'

'Okay whad'ya say you wash ma car for me 'n clean up the inside too, how'd that be?'

Off I'd go and an hour later JT would wander down the front steps of the office, his hands buried in his pockets rattling the change about. He'd take a minute to survey the silver Mercedes. 'Damn good job son.' In his big hand was a twenty-pound note. That was way more than any other neighbourhood kid got for washing a car.

On one occasion he let me drive his Mercedes to the shops and back. 'Don't tell your mom you drove, okay buddy?'

In the years Mum worked for JT he always took time to chat to me or walk around the workshop with me and 'bullshit to the boys'. Both JT and his wife June looked after their people in a way I have never seen again. They had us over for dinner many times. He had changed the course of our family's future for the best by employing my mother.

JT was still going offshore well into his seventies. I would have liked to thank him as an adult but I never took the time and sadly he died in 2002, the last of his generation. The oilfield is run by the corporate machine more than ever now, lawyers backed by engineers who have never seen a rig. The human side is gone, the bottom line rules, and it's every man for himself. Outside my crew I don't trust anyone. Not in the office, and especially not the client.

These days it leaves the older guys, who remember the rigs when they were still wild, seething. They speak up on occasion, usually at a critical moment during the meeting before the main meeting which proceeds the really big meeting where we talk about what we're going to say when we have the really really big meeting with the people in Houston joining us via satellite speaker phone (that's the meeting where no one makes a firm decision because of the consequences of getting sued for making a decision). 'Aw fuck all this horse shit' the old guys say, when the bureaucracy gets ridiculous and the legal implications of opening your mouth have you more concerned about losing your job than actually solving the problem. And for the tiniest of moments everyone in the room is reminded of the qualities that made these men pioneers when the drilling game was in its infancy.

A year after Mum and John got together, JT offered John a job. Then when the company decided to open an office and machine shop in Australia, he asked John if he would like to transfer there, with permanent residency sponsored by the company. John and my mother jumped at the opportunity. I was fifteen and again life changed dramatically for the better.

Perth was a great place to discover Australia. I loved going to the beach, I started surfing, everyone was so nice . . . so many girls wearing so much less than they do in Scotland.

The next three years were a total blur: I didn't do well at school, I was far too busy discovering my first girlfriend, first beer, first motorcycle, and the fact that everyone had a pool in the backyard. Unlike rainy Scotland, barbecues didn't start and end in the garage. Fruits and vegetables were big, days were long and hot, and neighbourhoods were large and well planned with wide streets and clean footpaths. We had a house with a huge backyard; Mum installed a spa in the middle of the back patio. She planted endless flowers and ferns that flourished around the steamy pool. Within a year our house and especially the back patio looked like a tropical oasis.

John got a big 4WD and the two of them went exploring on weekends, and often flew off to Asia on long weekends. For some reason they trusted me to 'look after things' while they were away. I was in teenage heaven. Just after my eighteenth birthday Mum and John got married. John asked me to be his best man; it was a special day, in a great new country.

I left school early, I wanted money, I wanted rigs, I wanted to fly in a helicopter, and say Gawd-damn a lot. I wanted to wear one of those gold company cigar rings. (What can I say, it was the 1980s.) But most of all I wanted an adventure. It's hard to get a rig job if you don't know someone, harder still if you're eighteen and green. I didn't want to ask John, I had to do it myself, but all I could get was a porter's position in a swanky hotel downtown. I'd answered an ad in the paper and they hired me. I was punctual and polite, but lost the job because I was caught having sex with a guest by a room service attendant.

A brief stint as a waiter followed but that too was doomed. The head chef was young; we would regularly hide in the back alley smoking joints and drinking wine. One night during the pre-Christmas rush of corporate dinners, we took things too far. Fran, one of the waitresses, was a law student who looked down her nose at having to wait tables. To rattle her cage, I hatched a plan involving the chef, the biggest tomato he could find in the cool room, some raw steak and a meat cleaver. Fran could not stand the sight of blood, and whenever one of the chefs nicked a finger she would run, covering her mouth and honking about hygiene and food. So I cut the tomato in half, pushed my white shirt cuff over my hand and dropped the tomato down inside the hole where my hand should have been. With some strips of steak hanging over the edges and lots of tomato sauce, it looked gruesomely real. Fran came through the double doors into the kitchen doing her balancing act, both arms full of dirty dishes. I spun around, waving my gory stump at her. She froze, turned white and made the long fall, straight down, face first into a banquet trolley of half-eaten food scraps.

Some years later I was living with three mates in a nice house not far from town. Pete, Phil, Iain and I got along perfectly well. There was never an argument or any of the standard scuffles you get with shared accommodation. The house was old but by no means run-down; we had a pool and a massive yard to grow drugs in. The back porch was huge, coming out of the house from two sliding doors then dropping down five feet to a surrounding rose bed that not only provided a natural balustrade of rose bushes but cover for our dope. Looking back now, it stands out clearly as the easiest time in my life.

Friday night was a ritual; it was my turn to fill the beer fridge out the back and provide dinner. We would sit around on the back porch in our underwear, everyone also wearing an obligatory funny hat, smoking joints and drinking beer, recalling the day's activities and planning the weekend.

One typically hot still Perth summer night when it was my turn to cook, I waited until everyone was sufficiently stoned and then called 'The Winged Wok'. Our Chinese takeaway food arrived and as we started peeling the cardboard lids off the steaming hot aluminium containers, someone farted. That may not seem overly funny, but when you've been smoking Pete's special Denmark Death blend it's bloody hilarious.

Iain roared, throwing himself backwards on the back legs of his plastic IKEA chair. But the legs snapped off, leaving him rocking precariously on the edge of the porch, so then he lunged forwards grabbing at the tablecloth, which dumped thirty dollars of piping-hot Chinese and a fruit bowl full of grass into his crotch. The high-pitched scream made us all grit our teeth as we watched him disappear over the side.

We peered down over the edge of the porch at him, sitting in amongst the roses which Pete's mother had pruned the previous weekend, covered in sweet-n-sour and dope. Iain didn't look up. We stopped laughing and jumped down to help him up, but he was frozen still, with his arm grasping his right thigh where the pruned stump of a rose bush was protruding through it.

'Oh fuck . . . call an ambulance.' I tried to scrape off the Chinese food and dope covering Iain's leg.

Phil grabbed the phone.

'Yes . . . he's in shock, I think,' Phil said, trying to sound normal. 'You better bring a hacksaw or something, okay, okay.'

'Where the fuck is Pete?' I yelled.

Pete came running up with his camera. 'Okay . . . put his hat back on.'

'You fuckhead!'

'I'm okay boys,' said Iain quite calmly. 'There's no pain.'

'That's 'cause you've been smoking this shit all night,' said Phil, scooping up a handful of greeny-orangey goop.

The camera flashed, making Iain blink.

'Someone's pulling into the drive.' Pete panicked. 'Hide the dope!' But his paranoid attempts at hiding the dope only had him smeared in it from head to toe.

The paramedics took one look at us and asked, 'Has the patient taken any drugs or alcohol . . . other than what he's lying in, gentlemen?'

In what seemed like seconds they had Iain on a gurney with the rose branch still through his thigh, sticking up under the sheet.

'He looks like he's got a boner,' said Pete.

'Can we come too?' asked Phil, but they slammed the ambulance doors and were gone, sirens blazing.

I suggested we follow them and so we all piled into my Holden, still in our underwear, liberally covered in Chinese food and marijuana.

'I'm still hungry.' Pete was eyeballing a McDonald's drive-through. So then we stocked up on burgers and munched our way to Fremantle Hospital, Pete dipping his chips in the greeny-orangey goop.

Not surprisingly the hospital staff wouldn't let us in but they did accept a burger, which Phil insisted they pass on to Iain as soon as he felt up to it.

2

POOL BALLS

1987–94

AFTER WRITING COUNTLESS LETTERS to drilling contractors I finally landed a job, as a roughneck (labourer) on a land rig drilling core samples for Western Mining. The job meant moving to a tiny town north-east of Perth, about twelve hours by car, but I was happy because I got what I wanted. So I packed up my old Holden, the boys wished me good luck, and with just enough money for petrol, I made for the rig.

Leinster is the middle of nowhere. A mining community of about ten thousand, it is 376 kilometres north of Kalgoorlie, the nearest large town and centre of Western Australia's mining industry. I was assigned a staff house on the edge of town, sharing with Craig, a geologist from Brisbane. Craig and I got on very well. He too had recently arrived and we found a lot of common ground to occupy our leisure time.

We decided to throw a party for my birthday, only about ten people showed up, but we had a good time. I drank too much scotch and tried to spot some hash with a hot knife over the stove, but ended up passing out on the kitchen floor. That's when two of the guys who were electricians decided to go out to their trucks and get as much gaffa tape as they could carry, then tape me naked to the kitchen floor. I came to thinking I was a paraplegic. Everyone spent the next hour laughing at my attempts to pull the tape off my body. We're not talking about a small piece of plaster here—by the time I was tape-free I didn't have any body hair left either.

Craig gave me a big joint 'for the pain', so I did manage to calm down, then half an hour later, I had the munchies like never before. But as usual we had no food. Luckily, the camp chef had made me a birthday cake earlier and had even saved me a slice during the feeding frenzy that occurred while I was taped to the floor. Of course, it was the best cake I had ever tasted but I was still starving. Looking into the fridge the way you do when you're stoned at two in the morning, hoping that by some miracle you're going to find a whole barbecued chicken hidden in the back, I saw nothing but beer and mouldy half-eaten TV dinners.

Then the camp chef told me that when he was making the cake, he had cracked some eggs into a tea cup, and suggested I fry them up for a sandwich. Like a man possessed I grabbed the tea cup from the fridge, threw the eggs into the big frying pan, lit the gas plate and started frying. But nothing was happening. The eggs wouldn't cook. I checked the gas, it was on; I turned it up. Still the eggs just sat there, refusing to cook. Hungry, half-naked and still sore from pulling every last hair from my body, I ran into the living room and complained, 'My eggs won't cook.'

By this time the party had reached stupor stage and no one felt inclined to help me. Craig, however, managed to haul himself off the couch and make the long trip to the kitchen. But when he got to the stove he started laughing. Craig had discovered why my eggs wouldn't cook. 'You're trying to fry apricot halves, Pauli.'

One of the guys at the party was called 'Riff'. Riff tried to play an electric guitar solo while standing on the coffee table, but it broke, sending him into the lap of our large neighbour who became disturbed and began punching him. Riff used his guitar as a shield and ran outside. I followed to calm him down and we talked in the driveway for a while. Riff was the town garbage man and needed an offsider to help him with his Saturday run. I needed the money so I applied and he agreed.

The next day Riff arrived with his flatbed truck. Walking up to the passenger door, I thought that my massive hangover was affecting my vision because I couldn't see him inside, but then I realised the whole cab was thick with smoke from the bongs he had chain-pulled on the way over. I got in, holding my breath, and wound down the window.

'Keep it up mate,' said Riff. 'Unless you want two million flies inside after the first stop.' He shoved the bong back in his mouth and flicked his lighter.

We drove to the first camp, where bulging, black plastic bin liners that had been sitting in the desert heat for days were lined up outside the camp kitchen porta-cabin. I pulled at the first of a dozen bags, but the knot at the top came off in my hand like melted cheese. Rancid garbage and thousands of blowfly maggots spilled down the sides.

'You're gonna have to shovel those into the truck mate,' yelled Riff from the cloud inside the cab.

So I tied my shirt around my head, leaving just a tiny slit to see through, and shovelled for the next forty minutes. Billions of flies followed my every move, deter-mined to get into my ears, nose, eyes, mouth—it was infuriating, not to mention disgusting. Riff did the next one, his long hair and stoned bliss making him oblivious to the marauding flies. And so the day went on, driving from camp to camp, one of us shovelling garbage while the other smoked bongs.

After the last camp we drove out to a remote site where a deep trench had been dug for the garbage to be dumped into, burned and covered up. By now I was quite stoned too and finding everything quite funny.

Riff backed up the truck to the edge of the trench, and I staggered out, climbed into the back and shovelled out the revolting mass. When I'd finished Riff hooked up the forty-four gallon drum of petrol and hand pump which he had tied to the back of the truck, then handed me the pump and climbed back into the cloudy cab. The pump wasn't working. I cranked the handle but no fuel came out, so I kicked the drum over on its side, letting all the petrol pour into the trench, then rolled the empty drum into the trench thinking, 'Job well done.'

Riff grinned at me through his dirty hair as I climbed back in the cab.

'I need a lighter,' I told him.

'I'll drive over there,' he said, raising a finger off the steering wheel pointing nowhere in particular. 'And you light the rag on this stick and chuck it in the trench, from a safe distance mind.' His eyes were so red he looked demonically possessed. He pulled another cone and handed me his lighter.

I jumped out with the stick and walked back to the trench, a black cloud of flies swarming around my head. Then following Riff's instructions, I lit the rag and threw the stick, but it fell short of the trench, teetering on the edge. So I wandered up and kicked it over.

All I remember is the sound, and getting punched in the whole body. The trench exploded in a fireball that people saw from the town miles away and the drum went into a low orbit. My shirt was on fire, and I started rolling around in the dirt screaming. Riff came running and helped put me out. My fringe was gone, so were my eyebrows, but I was okay. Then Riff was laughing. 'The whole fuckin' drum,' he kept saying, over and over again. We drove back to town, not a fly in sight.

The work was hard at Leinster but I enjoyed the vast open country. Time passed quickly. The rig was mounted on the back of a huge truck and I learnt the dos and don'ts fast. Des, the man in charge, treated his crew well and we respected him. But while work was hard, life's lessons were sometimes even harder.

On one occasion, I had a few days off during a blisteringly hot summer and Craig and I were driving down the main street when he suggested a beer in the Wet Mess, one of the two bars in town. Except this was the bar for the wild men.

'No way . . . I'll get raped,' I protested, but Craig was already getting out of the car and heading towards the bar, chuckling at me.

It was early afternoon so there were only four hallpack truck drivers shooting pool inside. They were all Maori; two of them had tribal tattoos covering one side of their face. All stood over six feet and looked like they'd been genetically engineered to crush small buildings. They nodded hello as we walked in. Everything was concrete—the bar, the stools—beer was served in plastic cups, and the windows had heavy bars instead of glass. We decided to play pool on the other table, and Craig paid the deposit for two cues and a tray of balls. (This was normal practice because every weekend they got demolished in a brawl.)

An hour rolled by and soon the four truck drivers started getting rowdy. They suddenly broke out in a vicious fist-fight, all four trading blisteringly hard punches. We panicked . . . but there was only one way out of the room, past them.

The fight spilled over on top of us. I made for the door but ran straight into one of the fighters, then his elbow ran into me. It was painless, really. I'd never been hit hard before, not hard like that. My brain went numb, lights out. I was on the floor, my nose wasn't working, tears were streaming down my cheeks. I could vaguely see the back of one giant bent over the pool table, his right arm swinging up and down, delivering his fist into the tabletop. I could feel the vibrations in the floor as he pounded on the felt. Then as quickly as it had started it was over, and they were gone. The barman was also gone, having locked the bar door and the steel grille between the roof and the bar counter behind him. My nose was smashed; blood flowed into my mouth and down the front of my shirt like two GT racing stripes. I got up slowly, and that's when I saw Craig. He looked dead.

Flat on his back in the middle of the pool table, he was covered in dark red blood, bubbles formed in the middle of his face. I didn't recognise him. The big trucker had shoved a ball in Craig's mouth, balled up another one in his fist and beaten Craig's face into a pulp. He had lost all his teeth, his jaw looked broken, as did his nose.

I carried him out to the car and struggled to get him inside. His blood spilled down my back as I positioned him. My head was spinning. I caught my reflection in the window—I looked like I'd just murdered someone.

The tyres shrieked to a stop in the doctor's driveway. Craig was unconscious, slumped forwards against the seatbelt, his head hanging down with a series of bloody saliva strings connecting his face to his crotch. A young woman was at the door, telling me that her father the doctor was in Perth for a wedding and she was unable to help me. The nearest medical help was a two-hour drive to Leonora. I ran back to the car.

Craig had a pulse, and was making a rhythmical gurgling sound so I knew he was sort of breathing. I floored the car as much as I could, regularly checking his pulse and trying to light bloody cigarettes with the car lighter. Finally I began passing signs to Leonora and felt triumphant just getting him there alive. The doctor lived in a modest whitewashed house and he had the flying doctors on final approach for the main street within an hour.

The plane's large rear doors swallowed up my friend

in superfast time, its departure sandblasting everyone in red prop wash as it vaporised down the main street and into the afternoon's dust-bowl sky. The doctor explained that Craig was stable but would need facial surgery and new teeth and his jaw was going to take more than a month to reset. He gave me a shot of anaesthetic, then he straightened my nose: shoving a wooden tongue depressor between my teeth and bracing my head between his knees, he quietly said, 'Now this is going to hurt.' It did.

During my drive to town, I had managed to cook the head gasket on the car, but luckily the doctor lived next door to a used-car yard so I just traded it for a Ford ute that had 'Killer Deal' painted across the top half of the cracked windscreen. I drove back to Leinster slowly, with my face a mess, bruised and swollen, squinting into the sunset on the straight desert highway through the 'Killer Deal' windscreen.

Six weeks later the four men had been fired over the incident and charged with grievous bodily harm, and all four were in Perth waiting for their day in court. Des asked me to drive to Kalgoorlie and pick up Craig at the hospital. He'd organised a new company truck and suggested that we go from the hospital to the car dealer, and from there Craig would drive the new truck back to Leinster. I decided to drive Craig's Toyota Hilux to get him; it was old and slow but more reliable than my 'Killer Deal' ute.

I left at daybreak and by early afternoon I was sitting in the hospital waiting for Craig. The doctor walked him into the room, one arm around his shoulder. Craig had lost a lot of weight but that was insignificant compared to his face. Four bolts protruded through the skin, all joined with tensioning wire that formed a square around his mouth. Anything he ate had to go through a blender first. I felt so sorry for him. He had to return in a month to have the bolts removed, but at least he could go back to work in the meantime.

I told him the plan, but he just wanted to have a beer. So we picked up the new truck and agreed to stop at the last pub on the way out of town. Craig didn't say much but he did have his jaw wired up, with a straw jammed in one side of his mouth and a cigarette in the other, so there wasn't much room for conversation. We stuck to two beers each—Craig sucked his—and as we walked out to the cars he unexpectedly turned and hugged me.

'Thanks mate,' he said and smiled with his eyes.

We had a convoy plan but minutes later I was crawling along in his crappy old Hilux and he was disappearing fast into the distance in the brand new Landcruiser. At least it was dusk, my favourite time in the bush.

It was easy to drift off in your mind on the long straight road, with ten minutes to the next corner and no radio, just the wind blowing hot against your skin. The desert stretched into the burning orange horizon in all directions. I felt lucky, I didn't miss the city at all. I paid more attention to the road as I took the solitary corner, but was still driving in a state of boredom. Heat haze distorted the road ahead, but I knew instantly that something didn't look right.

It was the new truck, lying on its side . . . I was looking at its black chassis. My heart jumped, I floored it, but the Toyota just blew smoke and grumbled at me. My mouth went bone dry as I pulled up next to the truck; I prepared myself for the worst.

There was a thumping sound coming from inside the cab. Then I saw a massive tail. It belonged to a wounded kangaroo, a huge red wounded kangaroo. Craig had hit him mid-hop and sent the beast through the windscreen into the cab. Somewhere underneath its bulk was my friend. It was kicking against the dashboard fighting for breath, and I had no chance of moving it. I didn't know if Craig was dead or alive—Do I go back to town? Get help? Do I try to move the roo?

I tried to get to Craig, but the kangaroo's tail was too strong and likely to thrash me into the broken windscreen. I couldn't see Craig, only angry kangaroo. So I went for the shotgun in the Hilux, but could only find one rat-shot shell in the glove box.

My hands were shaking so much I was worried I'd hit Craig, so I jammed the end of the barrel up to the roo's head and shot it point blank. But it was much heavier than me; I couldn't move it. Craig's feet were now visible so I sat on the road, one foot either side of the windscreen. With sweat stinging my eyes, the roo's blood tasting sweet in my mouth, one of Craig's feet under each of my armpits and using my legs for leverage, I pulled him through the windscreen and out from under the dead roo. He popped out onto the road, on his back; I had cut him on small bits of glass left in the windscreen frame. Scrambling up, I saw Craig's face and it filled me with horror; my mouth filled with vomit. The bolts that were around his mouth had caught in the kangaroo's hide and were now in the middle of his face, with big chunks of bloody fur stuck to them. Once again he was unrecognisable.

There was a pulse, however, and he was alive, so once more I strapped him in and drove back to town. The same doctor who had waved us goodbye was still on duty. He looked whiter than his coat when I came tearing into the hospital's main entrance covered in blood. Another six weeks later, Craig took the bus back. His good looks were gone but in his ever-positive style, life took on a new precious zest, even when he catches people staring. He just wasn't supposed to die young.

3

PACK-A-DAY MONKEY

1994–95

I SPENT A LITTLE more than a year working in Western Australia's goldfields when a friend of John's rang me out of the blue and offered me a job. The oilfields were booming then, with jobs available just about everywhere. I hopped from one company to another, working mostly in Asia. My twenties went by so fast I got whiplash.

In that first year working offshore my initial attempts at fitting in were fumbled. Then I got lucky and found myself standing on the drill floor with Erwin Herczeg. Erwin had done it all: run every kind of pipe, on every kind of rig, on three continents, in more than a dozen countries. His reputation was impressive to say the least, but he never bragged about it nor belittled anyone with 'been there done that'. As luck would have it, he took me under his wing and I learnt from the master. Erwin imparted his knowledge in a steady, patient way and I retained just enough to keep my limbs intact and my sanity preserved. I looked forward to any job that he was on, and I took every opportunity to work with him.

While my working life was on track, my social life became bizarre. When I got off a rig, I'd stand in front of the big board at Changi International Airport in Singapore and choose a flight to wherever. I had money burning a hole in my pocket and no financial sense, so I'd take off and fuck around in Tunisia for a month, returning broke to a rig with only some obscene Polaroids and one too many drunken stories.

After a couple of years rig-hopping around South-East Asia with different crews, I landed a job on Erwin's crew, based at Brunei. By now I had gained more experience and I really wanted to work with Erwin, but before I started in Brunei I took a break at home. Mum and John had some news. After nine years in Australia, they were being transferred to Songkhla, Thailand.

I dropped my mother at the airport, and Perth became a lonely place. That night I went into a new bar where a friend was working, ordered a beer and pondered. Miles away in my head, I thought about tracking down my father, perhaps spending some time with him because we hardly knew each other, I decided to make some calls the next day. Then I noticed a young woman behind the bar. She was fit looking, about five four, with short cropped black hair, bright red lipstick and a black T-shirt. I could tell straightaway that she was no ordinary woman. She had more attitude than most of the rig crews put together. I asked my friend who she was. Ruby turned her head and I called her over.

She had a great walk, purposeful, feminine. I was riveted. Somehow she managed to suggest all at once a mix of sexuality, combustible rage and poetic sensitivity. 'Hey,' she said and smiled, then someone else called her, but walking away she looked back just long enough to give me a 'Hey you . . . kiss my arse' kind of look.

I felt like I'd been shot in the heart. Unfortunately I was due to leave Perth in a few days so I promised myself I would get to know Ruby when I came back.

Brunei is a small sovereign state on the island of Borneo, located pretty much in the middle of the South China Sea. It's a beautiful place, free of most of the problems that corrupt other South-East Asian countries. The locals on the crew were hospitable beyond belief, hard working and cooperative without the ego-driven hardline attitude of the Western crews I was used to.

The staff house in Brunei was located in a small village on the coast. I shared with Drew, the base manager, who was a pleasure to live with, and our home life was clean, comfortable and quiet.

One day our neighbours came back from a trip into the jungle to visit relatives who still lived in an old-style 'long house'. They had a baby makak monkey with them. Someone had killed his mother with a blow gun and on retrieving the body discovered the infant still clinging to her. He looked pathetic, sitting in a bird cage on their front porch, just skin and bone and so small you could sit him in your hand. When Erwin, who regularly came through and stayed at the house, saw the monkey he took pity on the little creature, and before long the monkey was ours, acquired in exchange for some company caps and T-shirts.

We named him Joe and he quickly became a very cool pet. By the time he was fully grown, he stood at about one foot tall, with brown eyes and dark grey hair. For all intents and purposes he thought he was one of us. After his first year he developed a taste for beer, speed-metal music and headbutting the bathroom mirror. Unfortunately Joe also enjoyed the odd cigarette. This wasn't a worry at first, but then he started turning into a pack-a-day monkey, and, because he couldn't figure out how to light up, he would steal your lit cigarette, perch on top of a cupboard, coughing and smoking, and then discard the butt rather carelessly. We soon became very concerned about him burning down the house, especially if he'd had a few beers.

Joe only became a problem when he hit monkey puberty and started masturbating ten times a day. You'd be watching TV, glance over and there he was, on the couch, feverishly batting off through clenched teeth and a menagerie of high-speed facial expressions. He spent most of that time outdoors for obvious reasons. That pissed him off, so he took it out on the postman, and anyone else he didn't know who came to the house. At one point Joe got pretty bad, everything from verbal abuse up to, and including, throwing shit.

It was during Joe's puberty that the Shell drilling manager decided to drop by unannounced; he had never been to the staff house before and took us completely by surprise. A charming character, always keen to have a chat over a beer with anyone from a roughneck to the company owner, the drilling manager was one of those rare people with lots of power who knew how to handle it properly. On this day his wife came along too. She was typically Dutch, tall, blonde and stunning.

Brunei is a fairly strict Muslim country. Any women you see are always wearing a traditional 'baju kurung', and are totally covered up, exposing only their faces. I had been there for over a year, Drew three years, and Erwin had been in and out longer than anyone. We hadn't seen the female form in some time. Neither had Joe; in fact he had never seen a woman wearing Western clothing.

We'd been standing at the bar chatting for a while when Joe came in. He jumped up on the bar and stood level with Mrs Drilling Manager's breasts, his eyes like saucers, mouth open, staring from my chest to hers and then back to mine. In one lightning move he grabbed her right boob, just as she finished saying, 'Ooh what a lovely monkey, what's his name?' Then she was screaming. Her wineglass shattered on the floor as Joe deftly made his way up her arm onto her shoulder while she pirouetted around waving her arms as if she was being attacked by an invisible swarm of bees.

Sufficiently aroused, Joe leapt onto the ceiling fan and, doing about ten revolutions per minute, commenced masturbating. While Mrs Drilling Manager was readjusting her bra and regaining her composure, we tried everything to get him down—all kinds of tempting treats, Cuban cigars, my best single malt, the only copy of *Juggs* in circulation in the country. But the little bastard was on a mission. So we did the only thing left to do . . . we turned up the fan.

This particular fan was huge; it looked like someone had bolted a B-52 propeller to the ceiling. No one had ever set it higher than five on the dial, which was enough to pin three grown men to the floor, their cheeks rippling like skydivers. It turned into a test of Joe's determination: he was a tiny masturbating astronaut in a centrifuge, we hit ten on the dial, furniture began rattling across the floor . . .

'*Go for it little man*,' cried Drew. Joe was a blur of teeth and fur.

Finally he flew off, slamming into the far wall, unharmed but dizzy. I watched him really enjoy a smoke. The drilling manager was folded up laughing on the couch, his wife next to him with the molestor's little hand prints all over her top.

Joe's apposing thumbs allowed him to do a lot more than just fiddle with himself. He would regularly stand in front of the stereo, twisting dials and prodding buttons, and every now and again music would blast out, sending him wild. He would run off to look for his stick, come back and bash the stereo until the music stopped. Usually the music only stopped because we used the remote to turn off the stereo, but it was still a victory to Joe.

It took me six months to teach him to pee in the toilet and not hit the rim. This was finally done by getting him to stand on the front rim, facing the upright seat and leaning forwards so his hands rested on the cistern, thereby achieving the right angle, much like a human male does when trying to pee with an erection.

Joe and I had one of our greatest battles in the toilet. I had returned home from a job offshore to an empty house; Drew was on annual leave, Erwin was offshore. Joe had been home alone, though we had installed a dog door for him to come and go, and three days a week a lady from the village came to clean the house and feed him.

I was tired and filthy when I walked in. I dumped my bag in the laundry, stripped off and threw all my clothes in the washing machine, and proceeded naked straight to the bathroom. In a slow exhausted trance I closed the door, took a shower, shaved and then sat on the toilet and was thumbing through *Sports Illustrated* when Joe burst into the house through his dog door. He knew I was home and ran from room to room looking for me. I called out and in the inch gap between the tiled floor and the bottom of the door I could see his little feet standing on the other side. Joe put his head against the floor and stretched out one hand under the door. I dangled some toilet paper just beyond his reach and he tried to grab it. But our little game didn't last long.

Joe jumped up onto the door handle. The bathroom door was, I think, someone's front door at one time and had a big old-style lock and key made of heavy brass. Before I realised what was happening, Joe had turned the key and locked me in.

I had that same awful feeling you get when you're about to board a plane and remember you left the iron on. Joe started bashing the key against the door and chattering excitedly to himself. Realising how much trouble I was in, I jumped over to the gap at the bottom of the door and tried to coax him into giving me the key. But he wouldn't.

The door was made of solid timber and opened towards the inside, so I had no chance of breaking it down. There was no window, and no one was due to come to the house for another two days. I pulled the shower curtain off the rail and slid it under the door, hoping Joe would get tired of bashing the key against it and drop it on the curtain, then I could pull the curtain back and get the key. But no, he wandered off and left me sitting on the bathroom floor.

I visualised spending the next forty-eight hours there, crying myself to sleep in the bathtub wrapped in a towel. I could hear the guys saying, 'Grown man gets locked in the toilet for two days by a monkey, what an idiot.' I plotted revenge: 'I'm going to skin that little bastard and turn him into a toilet seat cover when I get out of here.'

After three hours of failed attempts at escape that included taking the heavy porcelain cover off the toilet cistern and bashing it against the door, and pulling a steel downpipe off the ceiling so I could start tunnelling—through the roof if necessary!—I finally figured it out. I straightened a metal shower-curtain ring and used it to knock out the pins in the door hinges. It took ages but eventually I got the pins out, removed the door and staggered into the hall.

Joe was sitting on the couch with the key next to him. He waited just long enough for me to see him then bolted outside, giving me a week to cool off before returning home, where the bathroom key was now nailed to the door.

Later that year the village started preparing for the annual 'Hari Raya' celebrations, but because it fell on the same day as the Chinese New Year, something that happens only every fifteen years, the party atmosphere was intensified. I decided to go over the border into Malaysia and get some beer, returning with enough alcoholic supplies to keep Erwin and myself amused for the evening, and a big box of firecrackers.

We sat on the porch, got drunk and lit the fire-crackers, which went off with a hell of a bang. I decided to blow up a coconut. The whole area around our house was littered with coconuts which fell from the palm trees, and I found that the older ones were perfect because the gap at the top where they had grown on the tree was big enough to push a firecracker inside. The resulting explosion was massive. I would light the fuse with a cigarette and bowl them down the dirt road where they vaporised in huge balls of white shrapnel—great fun.

Joe had been locked in the house but somehow got out. Just as I let go of a lit coconut, he passed me, running after it down the road. He thought it was a game. I took off after him and the bouncing coconut, but didn't close the gap in time. Joe jumped on it just as it went off—BANG—he flew off into the jungle. I screamed. We found him straightaway, but his fur was scorched, his body totally limp. I pressed him against my ear, listening for a pulse. He was gone.

I was devastated. I had stupidly blown up my monkey with a coconut, and we're supposed to be more advanced than they are? Have you ever tried explaining yourself to one of them?

The next day I buried him with his favourite toy, a can of his preferred beer, a pack of cigarettes and the bathroom key.

To get my mind off Joe's demise I was sent on a HUET (Helicopter Underwater Escape Training) course. Every two years all personnel who work on offshore installations have to go through HUET. It's a two-day intensive course in how to crash. Supposing you survive the impact, and let's face it, auto rotation aside, lift versus drag and rotation is all good until the rotors stop turning. If that happens a helicopter will drop like an anvil with a tractor tied to it. HUET is designed to imprint the correct egress method to save your life. It's also fun, and all done in a fantastic simulated environment.

There's a scale copy of a helicopter hanging over a giant pool, with wave machines, smoke machines, powerful fans, fake debris, everything you need for a good crash. The crew is strapped into the chopper with four-point harnesses, wearing inflatable life jackets equipped with lights, whistles, compressed air canisters, and of course your EPLT (Emergency Personal Locator Transmitter). If you're working in an area where the water temperature is cold, then as well as all the gear you have to wear a survival suit. That's basically a really thick, all-body condom that reeks of sweat and rubber.

The EPLT is a wonderful device, about the size of a pack of cigarettes, that transmits your position, accurate to within ten square metres on the global emergency frequency. So, should you find yourself bobbing about alone in the middle of the North Atlantic whistling 'I am Sailing' to torchlight, one of the most important bits of kit is your EPLT. I wish I had one for my car keys sometimes! Some years ago, an offshore worker finished a contract drilling job and stole his EPLT, taking it back to the United States as a keepsake. A few months later his son found it and activated it, and within the hour he had choppers and police cars swarming on his suburban home.

The compressed air canister, or spare air, is vital too. It gives you time to make your escape from the crashed chopper. Many helicopters are capable of making an emergency landing on water, like the Sikorsky SN series which has an underside that looks just like a boat, and most can inflate big pontoons to float and stabilise the aircraft in the water. However, helicopter engines are on their rooves, and depending on the sea conditions and how badly they crashed, they can overturn easily. And if that happens they sink, rather like an anvil with a tractor tied to it. I know of only three occasions where everyone got out unharmed from a submerged inverted helicopter that was ditched at sea.

HUET is designed so you can practise escaping from a rapidly submerging upside-down fuselage. They simulate all this very well: first the pilot calls 'BRACE' and you adopt the brace position, then you wait for the impact with the water, take a big deep breath as the fuselage fills up with water fast, stay still as it rolls over, and hey presto you're sitting there strapped in upside-down.

If you're next to a window, you pull out the rubber lining from the frame around the perspex window and punch it out with your elbow, then release your four-point harness, climb out the window, being careful not to snag your 200-pound beer gut on the edge, pull the cord on the life jacket, and float leisurely to the surface of the pool. If you are unable to hold your breath any longer, then by all means use the spare air canister. If your life jacket fails to inflate, then just follow your bubbles to the surface. You have to do it three times from different seats in the aircraft, so you get used to opening combinations of different hatches, doors and windows. Oh and all this can happen in the dark.

We did the HUET course once in Asia. The entire crew was hungover, and two guys swallowed so much pool water they vomited as we rolled over. I watched them disappear in the dim light into a cloud of barf. The guy next to me released his harness before he jettisoned his window and so lost any leverage he had to open it, instead opting to kick me in the head until he couldn't hold his breath any more. I just sat there upside-down with the spare air in my mouth. I could make out Ambu, who had inflated his life jacket inside the aircraft and was now pinned to the floor. To my right it looked like an impromptu rugby scrum, four guys were attempting to use one small door at the same time, and one of them was trying hard to suck air out of his radio transmitter. My air ran out just as they lifted the whole thing back out of the water, with everyone still inside and Ambu still trying to figure out how to deflate his life jacket. The giant pool was littered with bits of gear and vomit. Jack the HUET senior instructor and safety diver was always great and showed superb patience, but that day I think even he was tested. His relief when we finished HUET and moved on to sea survival, where we were joined by two pilots, was short lived. Every time they mentioned 'dangerous flying situations' we would yell out 'Hi-Jack' and wave at our instructor.

During my Brunei stint, I went back to Perth for a few months' time off. I rented a small house near the city, purchased a second-hand car, and put the jungle, the rig, Joe and Asia out of my mind, at least temporarily. Ruby was right where I left her. We talked and effortlessly fell into a fast friendship. But my romantic thoughts were shot down quickly. Ruby was direct: 'You're just not my type. Sorry mate.'

Over the next three years, however, Ruby and I became very tight. She could raise hell partying from Friday night 'til Monday morning, come home and go straight to work looking like the front cover of the latest 'single white female and proud of it' magazine, whereas I looked like I needed a heart–lung bypass just to get to the front door.

Living in the jungle of Brunei puts you in touch with life at its most primitive. It's not for the sensitive products of Western society. In the jungle, you have to kill something if you want to eat. In the jungle, there is always something trying to eat you. Competition is incredible. There are thousands of links on the food chain. I always thought the jungle would smell like boiled Kings Cross, but it doesn't. It smells great, and the jungle floor is always clean—because the moment something slows down it's eaten.

While I was in Brunei, the oil company was spending money on training and team building, positive activities for guys like me who were isolated for long periods of time. One of these team-building exercises for the upper management was held in the Brunei jungle. A few of us 'commoners' also got to be involved. It was fun explaining to upper management that there was no shower or cold beer. They thought it would be a piece of cake, like one of those 'outward bound' courses. You should have seen their faces drop when the team leader said, 'If you boys want to eat, you're going to have trap and kill something.'

The upper management guys were not used to the thought of having to kill for food. For most of them, hunting and gathering meant rolling down the car window and grabbing a burger, which they could do without too much trouble.

The upper management exchanged blank looks until finally one of them took charge. The idiot actually tried to lure a monkey from its branch with a fucking banana. Then he attempted to beat it to death with a rock that he cleverly hid behind his back. Of course, the monkey, with a lifetime of guerilla warfare experience, promptly retaliated by getting his mates to systematically piss all over the manager. Wherever he went it was open season, and for the next hour all you could hear was whooping and chattering from the canopy as the monkeys had a laugh at his expense.

At the end of the day everyone was shattered. Their beds were simple 'A' frame hammocks, slung a couple of feet off the ground. One guy was freaked about having to spend the night in the bush, so he popped a couple of sleeping tablets. His hammock sagged during the night and he woke up to discover that half the jungle had crawled and slithered into his shorts—even his bites had bites. He screamed like a madman, rolling on the ground and fishing madly in his crotch.

It was great to see these arrogant men who enjoy throwing their weight around in the business world so wonderfully far out of their depth. Standing around in the jungle, bunched up, paranoid and alienated, businessmen look as out of place as a 50-foot pyramid of severed heads in Taylor Square.

Surprisingly, most of the businessmen really enjoyed the jungle experience. They learned new things about themselves, like how not to beat a monkey to death, and dropped a few pounds in the process. Although, as soon as they got back to Singapore, it was beers and dinner and 'Thank God that's over'.

In the oil business, like most industries, it's the accountants and lawyers who call the shots, and these people make decisions that ultimately put crews in situations that affect lives in ways they could not possibly comprehend. How these team-building exercises were supposed to help them make better lawyers and accountants I don't really know. A wise man once said, 'The road to hell is paved with lawyers and accountants.'

(4)

SATURATION

1995–96

SHELL LAUNCHED A MASSIVE 'work over' campaign during my second year in Brunei. A work over is basically an existing producing well that needs a service. The rig simply retrieves the old pipe and runs a new pipe back in the hole (the completion string). Any special items on the string other than the pipe itself are referred to as 'jewellery'. These can be any number of things, from down-hole motors to mandrels and radioactive sources for survey purposes.

In the work over, part of the new 'string' was a down-hole titanium gauge. This was very expensive jewellery. Our one-of-a-kind-specially-made-don't-fuck-this-up titanium gauge was getting picked up off the supply vessel. Because the seas were a bit rough that morning the crane was using the whip line. But the whip line promptly snapped and there went the we-only-had-one-of-those-you're-in-big-trouble-now gauge into the sea.

Usually there is a back-up for every conceivable thing that could go wrong. But this jewellery was a one-off, and replacing it was not an option. There was simply no time to specially make a replacement gauge, let alone the mammoth cost of airfreighting a 30-foot long hunk of metal halfway around the world. This was going to be a retrieval, using a crane barge and a saturation dive crew.

Saturation diving is very dangerous; the men who do it are a breed apart, a group within the group on a rig. It's rare to see them on a rig these days as more oil companies are utilising ROVs instead. An ROV is a Remote-Operated Vehicle, basically a submersible with mechanical arms that can perform all manner of tasks on sub-sea equipment. The ROV is piloted from the rig via a cable tethered to the sub—it's like playing a really cool computer game.

I have met a few 'sat' divers and the things they told me raised the hair on the back of my neck. Most of the horror stories involved the hyperbaric chamber, where the crew goes to decompress after a job. The chamber sits on the deck of the rig, a cramped metal tube-like container with only one small round window. Its internal atmospheric pressure matches the depth pressure that the dive crew had been working at. They have to stay in there for days, as the nitrogen in their blood slowly escapes, totally relying on the support staff to bring them food, monitor their progress and take care of their lives.

One 'sat' diver came to grief in the toilet. He was unfamiliar with the old hyperbaric chamber they were using. He went to the toilet, located in the only separate section in the chamber, sealed the hatch and did his business. Flushing the loo in a hyperbaric chamber is a complicated affair, especially when you're knackered after a long job. You have to operate a number of levers and valves that seal the toilet and then the poo is sucked out into a container. This process happens in a fraction of a second—your poo equalising to atmospheric pressure instantly. This diver got the levers back to front and ended up with his bum creating a seal around the toilet while his insides were sucked out in a millisecond, killing him instantly. Another guy decompressed too fast and the expanding airspace between his fillings and his teeth had him rolling about in agony on the deck as his teeth exploded.

Our 'sat' divers arrived within twenty-four hours of the gauge going down, but in that time the weather had turned nasty and all operations were shut down until it was safe. The divers had been briefed on the job and knew what could and couldn't be done in Brunei. Although it's a Muslim country, expat workers are allowed to bring in one litre of alcohol each. The eight-man 'sat' crew followed procedures to a tee, but the weather was against them so they settled into the only decent hotel in the village to wait it out.

Two days went by with no change. Late at night on the second day the divers got really bored. There was no night life in the village, of course, so they each drank their allocated bottle of scotch and made their own fun. When a 'sat' crew is on stand-by, and there's nowhere to go, no women to chase or bars to demolish, the make-your-own-fun scenario usually ends badly. These include some of the more amusing near-death by misadventure stories you hear. For our boys Brunei was not a fun-rich environment, with a distinct lack of things or people to fuck with, so they went for the biggest thing in town . . . the mosque.

Boltcutters easily breached the main gates. Access into the main building, via an open window, was even easier. At the top of the big tower in the middle, they found what they were looking for. Years ago the Koran was belted out from the tower via a megaphone and an open book, but these have been replaced with a time-delay tape deck. The prize sighted, the divers exchanged tapes. They also changed all the padlocks they could find on their way out with the locks from their offshore kit bags. The 5 a.m. call to prayer was not a good one; most of the village was head-down-arse-up as Johnny Cash's 'Burnin' Ring of Fire' wailed over the rooftops. The religious police got involved, and it took hours for the locals to get in and turn off the tape. The divers were lucky they only got kicked out of the country with 'Religious Offender' stamped over their passport photos. We never got our gauge back.

I went to see my parents in Thailand and had a great time. Mum showed me just about every side of Songkhla from the best local restaurants, hotels and temples, to umbrella factories and the boot camp where they take teenage girls from the kampong and teach them how to shoot a bizarre array of inanimate objects out of their genitals, including bananas, ping-pong balls, even a blowgun. They can open bottles, smoke a cigarette, draw a picture, solve a Rubik's cube, all with the holiest of holies. It's a sad place, where innocence is burned alive for a quick profit from a drunk tourist.

Mum was going to the orphanage on the edge of town twice a week to teach the kids English and she took me into the building where they try to look after the HIV-positive babies. Born into death, they didn't stand a chance. It shook me like never before and tears ran down my cheeks—we're useless to them. I walked out ashamed of my ignorance, my easy life, and how lucky I have been.

I also experienced 'Songkran', the Thai water festival. I had no idea it was a special day as I walked to the market.

An old woman opened her front door, babbled something at me and hurled a bucket of water at my face. So I walked home, completely bamboozled, changed into dry clothes and re-emerged on the main street, only to get hammered with water bombs by a truckload of teenage boys. Home again I learnt about 'Songkran'—the Thai new year celebration where people throw water at each other as a symbol of purification—and returned to the main street prepared, with John's large fire extinguisher in my big backpack. Looking like a wannabe 'Ghostbuster', I was ready to join the fun.

One morning I woke to the screams of my mother and ran to the kitchen in my underwear. She had seen a big snake go under the fridge, but she told me not to worry as John had made a 'snake stick'. This was a two-metre long steel pipe, about an inch in diameter, with a sharpened U-shaped prong welded to the end. With this high-tech device I was somehow supposed to get said snake out from under the fridge and onto the front lawn, where it would remain perfectly still while I cut its head off.

I took the stick from my mother who said she would be upstairs phoning John in case I stuffed up. I stuck the stick under the fridge and poked about towards the back. The two-metre cobra, who was perfectly cool and happy under there, took the stick away from me and chased me out to the front lawn, where I'm sure he hoped I would remain perfectly still so he could cut my head off.

Then Mum screamed from the upstairs balcony, 'Oh my God Pauli, there's another one behind you.' The cobra's mate that no one had noticed was making her way around behind me.

They had trapped me. With nowhere left to go, I climbed up the 'Spirit House' in the centre of the front lawn. About two metres high, the Spirit House is a small model temple that most Thais have in the garden. Every morning, a glass of water and some food and sometimes flowers are placed by the Spirit House as an offering to Buddha. I knocked all that stuff off as I scrambled up. The cobras circled for a bit then began climbing too.

Thankfully John's foreman arrived in a truck. He took one look at me, perched on the Spirit House in my undies, then casually walked over, picked up both snakes, smiled at me and put them in a bag, waving as he drove off. I climbed down and spent the rest of the day recovering upstairs with Mum.

For the next twelve months I fluctuated between the 'work over campaign' offshore and the new well getting drilled deep in the jungle. Each offered up its own unique issues, but never got remotely dull or repetitive.

One hot and humid morning during the monsoon season in Brunei, we got the call to go immediately to the heliport and catch a flight to a land rig deep in the rainforest. We were going out to replace a crew who had been there for way too long. There is supposed to be an industry cut-off point for time spent on remote closed-in locations like that, but I have yet to see that implemented properly. This crew had broken all the records; they had been there for months.

Anyone who has spent a significant amount of time in the jungle will eventually either love it, and not want to leave, or hate it, and end up flipping out. On this occasion the crew consisted of locals who were completely at home in the jungle, so they were fine. The crew chief, on the other hand, was a Texan on his first job in South-East Asia, which is quite an adjustment, let alone the jungle. I was told on leaving the base that he had stopped talking a few days previously, and had spent the last twenty-four hours locked in his cabin.

The jungle can suffocate your mind. Over time a man can start to slip mentally without realising it. I had to replace the Texan immediately before he degenerated to the point when the medic had to give him a shot, put him in a straightjacket and send him home where he would be told to sit up straight, suck the drool back in and try to get a new job. I was looking forward to getting out there—I always felt completely safe in the jungle, much more so than on a Sydney street—and most of the previous six months I had spent offshore, so the prospect of working in the bush again was exciting.

The land rig was 150 kilometres north of our base, and the pilot was planning to shut down and stay there for an hour to refuel and get something to eat. That would give me enough time to talk to my disturbed American colleague and get him safely on the chopper for home. My boss's departing words were, 'Make it quick, and no fucking about okay?' And off we went.

Life is fragile enough without the occasional hint of death that lands in your lap. Enough time had passed since my last hint and I had lapsed back on my laurels, confident in the illusion that I was indestructible. My illusion dissolved just after take-off. It was only supposed to take forty minutes, over the jungle, down a valley, over a small river, and there it was, a hole in the canopy, a green tube illuminated by floodlights, with lots of noise, Diet Coke and cigarettes. I was already there in my mind, telling stupid jokes and catching up with the boys, but no.

Within ten minutes of take-off our chopper, a twelve-seater Sikorsky, was enveloped in dense low cloud, and it was obvious that there were not going to be any spectacular panoramas that day. So I talked to the crew, or rather screamed at the crew over the noise of the turbine, and tried not to notice the violent turbulence or succumb to my once well-hidden paranoid fantasies that had us all screaming as the rotor blades shattered and we began to make that fall into the abyss. I have always hated old choppers. I'm not fond of most choppers but old bouncy choppers really shit me to tears. They carry too few personnel to attract more than a passing nod when they crash, as they do quite regularly.

Every time I read *Upstream*, an oilfield newspaper, there's an article like this:

> Bumfuck Nowhere: all nine passengers and crew died yesterday when a twelve-seater Sikorsky helicopter operated by Doom Air crashed in a really big ball of flames shortly after take-off from Bumfuck Nowhere regional airport. Witnesses said the helicopter fell for, oh wow, ages before vaporising into the jungle at 1592 miles an hour.

So I just sat there and went to a happier place in my mind, which progressed into lurid thoughts of naked cheerleaders playing with a giant beach ball . . . Perhaps that had more to do with the way a chopper vibrates. I kept looking out the hole where the window should have been into the murk, when suddenly the aircraft shook and threw us all over the place as it became caught in a massive monsoon storm. These can appear out of nowhere in the jungle. Hot steamy air races across the jungle canopy and collides with cold offshore sea air, and a very big storm ensues. I was getting worried and could see the same look on all the faces around me; we were flying through one of those 'Dr Frankenstein . . . it's time' electrical storms. Lightning cracked down through the rain, vaporising all the moisture around us, the sonic boom making everyone jump in unison.

The pilot tried to drop out of the weather, falling fast for a few hundred feet into clearer air. When we were only about fifty feet above the tree tops, skimming over the canopy, we rose up sharply and started popping in and out of the low cloud like a confused pigeon. This went on for fifteen minutes, up and down, up and down. Every time we popped into clear air there was nothing but jungle as far as you could see. The pilot came on the speakers in our headphones, and in a calm Louisiana drawl said:

'I don't know if you boys have noticed, but I'm having a little trouble eyeballing the rig. We have lost some instruments and need to do dis approach by line-of-sight, and I've got a master caution light on my fuel, so I would appreciate it if y'all could yell out if you see da rig, okay fellas?'

Before he had finished speaking all eight men were craning their necks in all directions, desperately looking for the rig, but it was like trying to find a pin in a sea of broccoli.

Many of the chopper pilots who work in the oil business flew in the Vietnam War and possess a kind of 'dynamic lethargy' that makes them very calming to fly with. Our pilot radioed the rig and asked them to release red smoke, which allowed us to vector in on its exact position, touch down and disembark. We arrived to a muddy location and a disturbed crew chief. I was just happy knowing I didn't have to make the flight back.

The crew chief, John, was indeed locked in his cabin. I banged on the door and eventually I heard the key turn in the lock. I kicked off my muddy boots, put them on a piece of cardboard by the door and stepped inside. It was a standard portacabin, twenty by twelve feet, with two small beds at either end, two lockers, and a table and chair between them. It was in a real mess; rubbish everywhere, dirty plates and empty cans littering the floor. There was no window; it was dark inside, the only light coming from a lamp that was angled right down an inch from the desk. Big moths whizzed in and out of view, occasionally crashing into the desk lamp. I tried the main overhead light but there was no bulb.

John was sitting on his bunk in his underpants, just staring at the floor. He had not shaved in a few days and looked like he needed a good meal. I told him to get dressed and start packing up his gear. Walking over to the desk, I saw what looked like little piles of insect body parts, wings and heads in one, legs and mush in the other, and a razor blade and tweezers sat under the desk light.

'John, you're on this chopper mate. It's going as soon as it's refuelled and they've done some minor repairs,' I lied.

He just looked up at me and grunted 'Okay'.

I sat on the bed opposite him, wondering where to start cleaning up, as it was now my cabin. John was getting his gear slowly, pulling his towel off the handrail by his bed. Then I sensed movement by my feet.

Looking down directly between my bootless feet in the dark, I focused on what looked like a very hairy human hand crawling towards my right foot. It was the biggest spider I had ever seen.

Adrenalin shot through my body, and soon I was airborne grabbing the first thing I saw—my boot. This thing was a monster, the size of a cricket ball with legs, so I proceeded to bash it to death with my safety boot. Whereupon John proceeded to bash me over the head with his safety boot. Apparently, it was his pet, and John had been locked up in his cabin at night catching bugs and feeding this thing for weeks. He named it 'Walter' and he and Walter had become firm friends.

Most of Walter was now decorating the floor and the heel of my boot. John and I lay on the floor, he was crying and my ears were ringing and blood was running down my forehead. He was still crying when the medic sedated him and put him on the chopper. The storm had blown itself out. I watched the chopper disappear through the green tube in the jungle canopy, my head throbbing in time with the rotor blades, but it would heal, unlike John's.

5

KILLER MOUSE

1996–97

WE SETTLED IN AND started the job.

I loved being back in the jungle. I loved the smells and the sounds, and its intense green presence. At night floodlights illuminated the site attracting every type of nocturnal creature. During the monsoon season moths the size of dinner plates would whiz around the rig doing acrobatics that occasionally ended with someone catching one in the head, the impact knocking off their hard hat and leaving them looking like they'd just come off badly in a custard pie fight.

Monkeys would get braver every day, eventually hanging around the rig like groupies after a concert. I would leave my former-little-shop-of-insect-horrors cabin and pass a dozen monkeys all eyeballing me. 'Hey buddy . . .

got a smoke?' they would chatter. I would have a pocket full of nuts, always something, which I'd throw to gain safe passage, remembering from my experiences with Joe that pissing them off was dangerous.

All the guys on my crew were locals. My derrickman, Ambu, is an Iban, a descendant of the headhunters who originally ruled the jungles of Borneo. His grandfather lopped off his fair share of heads during the Japanese occupation in the Second World War. Ambu has the 'bamboo' tattoos of a headhunter around his throat, but says he only took a few heads in his forty years in the jungle. I always had time for Ambu. His ability to work in any condition made him invaluable on a job. He is not afraid of anything because he has his power. His power is a thin leather belt decorated with teeth and charms made from lead fishing weights. As long as Ambu wears the belt, he cannot die. Remember *The Lone Ranger*? Well, talking to Ambu was like having a conversation with Tonto.

'Come . . . we go.'

'Come . . . we eat.'

'Ambu . . . cannot die.'

Born and raised in the jungle, Ambu possesses the intimate knowledge that only a lifetime of living there can give you. Nothing in the jungle follows the rules as we understand them. Dogs don't chase cats, cats don't chase mice. Monkeys don't ask for bananas, they want cigarettes. Ambu, for example, arrived at the workshop in the village once with two dogs in tow. One was a big shaggy dopey-looking thing with a perpetual drool problem and the other was a small scruffy multicoloured guy who walked under the bigger dog. I asked if they were his dogs. Ambu pointed at the big one and said, 'She Kuching . . . She my dog . . . The other one is Kuching's dog . . . His name Arnap.'

'Your dog has a dog?' I asked.

Ambu nodded. 'She bring him home one day.'

It was on this job that I got my first taste of Ambu's amazing skills in the jungle and his ability to bullshit as well. We were shut down, waiting for another service company to fix an equipment failure. Everyone was bored brainless and sitting around in a small clearing at the edge of the site. Ambu pipes up, 'I can make the mouse kill the scorpion.'

Ambu was known for his little statements, so I said, 'Okay Ambu, off you go.'

'You wait . . . I bring to you.'

'I'm coming with you,' I said.

'Come . . . we go.'

He was excited to have me tag along and promised to show me something only a few white men have seen. I had to bring a roll of gaffa tape, a flashlight, a painter's mask, my sneakers, a small wood saw, goggles, gloves and rags. Stuffing all this and some water into a backpack, I grabbed a walkie-talkie from the radio room and we set off.

Ambu took off like someone does in the supermarket when they can't find the bread. I was lost after five minutes. I could hear the rig, but the jungle was so dense I had no idea where it was. The canopy blocks out a lot of light but that didn't stop Ambu from moving fast between the trees. Climbing around the rocketship fenders of a massive moss-covered tree, I found a beaming Ambu. He grabbed the backpack and pulled out the saw then plunged it into the centre of the trunk, as if stabbing an enemy with a sword, turning his head to spread a betel-nut smeared grin at me. That's impossible, I thought, the tree was not a tree, but something else entirely. I looked more closely. It had once been a tree, a long time ago, until a strangler vine crept up and slowly but tightly coiled itself around the trunk, spiralling from the base all the way up to the canopy. Over the years, the vine, like a vegetarian python, had throttled the life out of the tree and eventually the tree rotted away to nothing, leaving a hollow tightly coiled rope tube to the sky. Ambu cut a neat hole in the vine, tipped the contents of the backpack on the floor, pointed at them and said, 'Put on.'

The gloves were taped to my sleeves, the rags were wrapped around my head and taped to the mask and goggles, my collar was taped to the rags around my head, and everything was taped to everything else. With the flashlight in my top pocket, I climbed in. The darkness made it easier to cope with the slimy insect-riddled walls, but it was still revolting. My goggles steamed up quickly so I took them off, and by the time I was halfway up sweat had soaked through everything and I was wet down to my undies. When it got too tight to keep going, I pulled the saw from my belt and cut my way through. Finally I was at the top and able to stick my upper body out.

At first all I could do was pull off the rags to get some air onto my face. I felt instant relief, like letting go of your end of a fridge on house-moving day. I was sitting in the jungle canopy on the roof of a world untouched by man. It was breathtaking. I felt like I did when I walked into the Sistine Chapel and looked up at Michelangelo's panels. I was awestruck, my senses over-loaded with the beauty of it, despite the vines digging into my bum and the bizarre crawling monsters spewing out of the vine tube. You see the jungle is the wrong way around. All the things that make plants grow and help sustain life in the jungle, from slime to great apes, come from the canopy. It's so dense that it traps all the sunlight, water, everything. If you want bugs at home you kick over a rock, here you climb a tree.

Ambu was shouting something from below. I had been up there too long; Ambu had already found his scorpion and was eager to go and find the mouse. The climb down the vine was much more fun than climbing up—just imagine getting sucked down a giant vacuum cleaner hose.

I was too tired to go mouse hunting so Ambu walked me back to the rig and went on his own. It took him until the following evening to find the right mouse. We were sitting in the clearing, smoking and drinking coffee, when he turned up with a metal garbage can lid, some tongs from the kitchen and a metal ammo box. He made a little circle of rocks in the centre of our clearing, up-ended the bin lid and placed it on top of the rocks. We all fell silent and crouched down to watch the show.

Ambu flipped open the ammo box lid and using the tongs pulled out a big black scorpion. Everyone backed up a few feet, it looked so evil. The scorpion was placed in the centre of the bin lid, its pincers were raised and its curved tail, with the poisonous stinger, hovered over its body. Big enough to cover your hand, it could kill a man in a few minutes. Then from a small cardboard box in his pocket Ambu produced a tiny mouse and dangled it by the tail over the scorpion. It was just a little ball of fuzz with a pink tail, no bigger than a golf ball.

'So you're saying that puny thing is gonna kill the fuckin' scorpion?' one of the boys asked.

Ambu nodded and waited for the boys to start placing bets.

The kitty was up around a hundred bucks when he dropped the mouse in the lid. The scorpion went for it, but it was like a forklift truck and couldn't turn fast enough to grab the mouse, who just ran around the scorpion in ever-decreasing circles until it was directly behind the tail. The scorpion could only turn in its own space, it just wasn't fast enough. The mouse ran up the scorpion's tail, and, hanging on to it, started biting through the tip. In less than a minute, the stinger and poison sac were bitten off, then the mouse ran down the scorpion's back and bored its teeth into the scorpion's head. The mouse hung on, staying out of the pincers' reach, until the scorpion lay dead.

Ambu collected his money, then announced, 'You want to see the scorpion kill itself?'

We were all mesmerised by then. 'Yeah sure Ambu.'

So he lit a small fire under the lid and tossed another scorpion in. As the heat slowly started cooking the poor thing alive, it could no longer alternate legs to stand on, and speared itself in the belly, dying instantly. I had no idea there was a creature that given no choice would kill itself.

The next trip home to Perth became my last. Ruby had decided to move to Sydney and asked if I'd go with her to help her find a flat and get settled in. I was happy to go as I had never visited the east coast before and a few weeks later we arrived in Sydney.

It didn't take long for Ruby to find a flat she could afford and she also found work in the first week, pouring beers in a bar in town. I was due back in Brunei in a month, which gave me time to look around. Sydney was a change of pace from Perth, and I loved it. I decided to make it my new home on my next crew change.

I looked up my friend Barry who pushed tools on a Brunei rig and was home having some time off. He introduced me to his sister Louise, a successful business-woman running her own advertising agency, who was as much fun as her brother without the spontaneous loud outbursts. It was then that Louise first raised the possibility of being creative for a living. She saw my paintings; I had been painting for years in my spare time, but never showed them to anyone. She loved them and invited me to her office to meet John, the company's creative director. We all got along so well that whenever I was back in Sydney I would drop in to the office and talk with Louise and John. Eventually, over about twelve months, this turned into work. For an hourly rate I would sit with John in the studio and brainstorm concepts, think up 'tag lines' for print advertisements and write 'copy' for all kinds of things. I loved advertising, the whole process, from an initial idea to walking down the street or opening a magazine, sometimes months later, and seeing something you worked on. It gave me a lot of satisfaction.

For me Sydney can be a cage with golden bars; it's

easy to fall into a complacent stupor. I've even caught myself becoming interested in architecture! If I'm not careful I will soon have the mind of a backpacker. In Sydney, if you can get past the day-to-day living stuff— you know, can't find a parking space I can competently drive into, there's no credit left on my phone, and the dog's just been sick on the dash—life is easy, especially after spending months on a rig.

I did miss Perth for a while, but when I returned to finish packing up all my belongings to send over to Sydney I saw Perth differently. It was close to Christmas, and deserted, because everyone had gone somewhere else for the holidays. It looked just like a clean giant version of Bondi with no people. I spent the whole time playing 'spot the locals'. No people, no cars, no one open for business. Preparing for the holidays in Perth is rather like sorting out the household after a thermo-nuclear weapon has gone off. You're going to need everything from stockpiles of petrol to one thousand rolls of toilet paper.

I hired what seemed like the only vehicle left in Western Australia and thoroughly enjoyed driving down Perth's deserted streets. On those rare occasions when someone did pull up behind me and the lights turned green, they would patiently sit there and wait for me to leisurely push in the clutch, slide the stick into first gear and peel off without a care in the world. In Sydney you'd better be riding the clutch and completely in sync with the traffic lights so that your car is already moving as they turn green. If you're not, the bastard in the car behind you will be battering the horn and spitting all over the windscreen as he screams his pre-emptive road rage verbal attack.

During my last week in Sydney it rained constantly, turning Ruby's small street into a river. One night Louise was having a party at her house in Balmain, following a dinner in a harbourside restaurant. I was having an absolute ball. She really knew how to entertain friends and family, and it made me not want to go back to the jungle village and deal with the boys.

At the end of the night I jumped in a cab and set off through the rain back to Ruby's flat, near Kings Cross. The flat was on the top floor of a renovated Victorian two-storey building. The tin roof, however, was as old as the rest of the external structure and had rusted through. I was sitting on the toilet, reading the sports page from the weekend paper, when I heard a loud crack.

During that rainy week, water had accumulated in the space between the rusty tin roof and the renovated ceiling. I looked up, registered what was about to happen, and immediately started weighing up the odds of wiping first and running . . . or just clenching and running. I made a grab at the toilet roll but the whole ceiling split through the middle, dumping what felt like tonnes of cold water and gyprock on my head. Ruby just laughed when she got home. We had to use an umbrella to go to the toilet for the next few days.

A week later I was standing on the drill floor in the pouring rain telling the story to Erwin; Ambu came up to relieve the derrickman.

'I go up stair now,' he said and grinned.

The derrickman is the guy who works up in the big tower that juts out of every rig, rather like an industrial Eiffel Tower. It's a tough job in bad weather and inherently dangerous. Working in the derrick is all about timing; everything that the driller does affects what everyone else on the drill floor does, especially the derrickman as he's got nowhere to run. His only escape device is a static line tethered from the top of the derrick to the main deck below. Usually set at a tragic angle, this line, called a 'Geronimo' or a 'Tinkerbell line', has a handle that you grab and hold onto as you ride the line down to safety, controlling your rate of descent by moving the handle braking lever.

While he's up there, one of the derrickman's jobs is to 'stab' the pipe. This routine process involves the roughnecks on the drill floor lining up the pipe at waist level with the last joint of pipe sticking out of the rotary table. They need the derrickman to help line up the pipe so it's straight from the top end, therefore enabling them to screw the threads together. The derrickman releases the 'stabbing board', a pivoting plank like a diving board, then walks out to the very end and, wearing a safety harness called a 'belly buster', leans out at a horrific angle to make sure the pipe is lined up straight. So basically the derrickman spends a fair bit of time dangling off a board ninety feet above the drill floor. Having a good head for heights is a bonus.

The rain was coming in hard, but Ambu was unfazed as he made his way up to the derrick. The derrickman whose shift was finishing, Jake, is as funny as Ambu and comes from the same part of Borneo. The two of them have been working together for twenty years. He also speaks English like Tonto.

Jake arrived on the drill floor looking pissed off.

'What's up?' asked Erwin.

'Aw fuck . . . I lose my teeth,' Jake was pointing at his mouth. He chews tobacco constantly, which involves a fair amount of spitting, and while leaning out on the stabbing board he lost his false teeth down the pipe. That made three sets so far for the year, the last pair going down the toilet when Jake got seasick during a storm on our last offshore job. It looked like we were in for another set.

The wind picked up a few hours later, and horizontal rain lashed the drill floor. The difference between rain at home and rain in the jungle is the difference between shaking off your brollie and looking like you just got flushed down the toilet. But nothing short of a typhoon will stop the job. The sky flashed as if taking a giant photo, and we all blinked as the thunder cracked down on our heads.

We had a couple of hundred joints still to run in the hole, and all I could think about was the end of my shift and a warm bed to crash in. Another flash and bang . . . lightning hit the derrick. Everything shorted out, followed by a few seconds of darkness and then the emergency battery lights came on. The driller gathered everyone; we all looked okay. I ran over to the intercom, thumbing the talk button.

'Ambu . . . Ambu . . .'

I ran into the middle of the drill floor and craned my neck back, scanning the inside of the derrick . . . but I saw nothing.

'Get a man in a riding belt up there now and another up the ladder.'

Then we heard him coming down in the tiny emergency elevator. Everyone froze and watched its slow descent to the drill floor.

The door flew open and out stepped Ambu, his hair standing straight up. He had one hand in his mouth.

'PPPPaul . . . My teeth are hot . . . My teeth are hot.'

I pulled off a glove and put the tip of my finger on one of his teeth. His fillings were hot, and so was the zipper on his coveralls. For once, he looked really frightened.

'Okay . . . Go and have the medic check you out. Ambu, got your belt on mate?'

Ambu was not hurt because he had his power belt on, at least that's the way he tells it.

After the jungle stint and we all got back to the base in one piece, I took a break, opting to go and visit my father. It had been a long time, and I was a little nervous about seeing him again.

The drive from our village to the capital of Brunei was a smooth journey for the most part. At the time, the Sultan was constructing a freeway that was to cut through the jungle, connecting all the small villages dotted from the southern-most point to the capital Bandar Seri Begawan in the north. Eventually there would be new sealed roads from every village, all flowing into the main artery like a concrete Amazon River. But for the moment I was driving on a potholed single-lane road that was losing its battle against the jungle which threatened to swallow it.

The staff car was a clapped-out fifteen-year-old Mazda that struggled to sit on eighty, but I was in no hurry. Glancing in the rearview mirror, I saw an ancient Greyhound bus looming up on my tail. I moved over to the left so the driver could pass safely, and the bus began to overtake me, kicking up dust and diesel fumes into my open window.

I was about to start cranking up the window when I heard what sounded like the biggest belch ever. Looking directly at me from only a few feet away was a buffalo. He had been unceremoniously hog-tied, up-ended and slid on his side into the luggage bay of the bus between the front and rear wheels, with two of the sliding doors left up as his head was too big to fit in. We looked at one another for a moment, his big eyeball registering mine, and then he was gone.

6

THE DEVIL'S BUSINESS

1997–98

TWENTY-FOUR HOURS LATER I was on the London Underground going from the airport to the city where I was hooking up with Steve, an old friend, in Leicester Square to talk rigs and life, love and failed relationships— poor man's therapy with too much beer and a hot bacon and eggs vindaloo by morning. The Tube is always crowded and noisy; Londoners, much like Parisians or New Yorkers, are tolerant of their personal space being invaded daily. On the other hand I was totally uncomfortable, having arrived from a lush wild land where just seeing more than a dozen Westerners in a week was extraordinary. In the middle of a typical smelly tunnel, the rattle of the overcrowded peak-hour train stopped. The lights flickered and went out, a few mumbled complaints wafted around. How British, I thought. Anywhere else and someone would be lying stabbed on the floor. The train driver's voice came over the speakers. 'London Underground would like to apologise . . . for everything.'

Steve has an odd way of looking at the world, but he's also fun. I asked him how he copes with being thrust back into his inner-London 'normal' life after living on a rig for months, often in a rough country. He pondered it for a few seconds then replied in his Cockney accent, 'Well me old son, I get owme from Eathrow . . . Straight down the boozer . . . Get abowt ten pints-o-lager down me neck 'n go down Piccadilly at five in the mornin' wiv a packet of bird seed 'n a cricket bat . . . I fackin' ate pigeons . . . no wot I mean?'

The next day I was on a platform at Paddington station waiting for a train to my father's place. I was hungover from my evening with Steve, and had the worst case of jitters since that chopper got lost in the jungle. I hadn't seen my father for many years.

To my initial horror I saw him do a whole menagerie of things that I do. His mannerisms, his voice, the way he walks when he's pissed. If you haven't spent any time with your father and then suddenly spend a week with him as an adult, it's confronting. But we had a fantastic time and at least I know what's going to happen to me when I get older. If I make it to old age, I'll check out the way I checked in . . . fat, bald and dressed badly, with a mild boob fixation.

During my Brunei stint I found myself back in London, en route to Aberdeen and the North Sea. A job had come up out of the blue, and I would have been a fool to refuse.

The North Sea is an oilfield icon, the centre of the drilling industry where historic breakthroughs have occurred, as well as some of its worst disasters. It is one of the roughest seas on Earth, but is also capable of producing six million barrels a day. Since its initial boom with the 'Forties Field' in the 1970s, it has become a sea-borne exercise in maritime gridlock. Its unbelievably crowded waters, sprouting platforms in all directions, ebb and flow on a mammoth scale and it has become a white-knuckle obstacle course of freighters, rigs, tankers and commercial fishing boats.

I was standing by in a hotel in Aberdeen, but a two-day wait turned into a four-day wait and on the fifth day the job was cancelled; the North Sea was not to be for the moment. Hanging around had given me a chance to look up some old friends so it wasn't a total disappointment, and I had a day in London on the way back to Brunei.

Steve was away, which was probably a good thing, so I decided to have a look at one of London's lesser-known evils. This city has produced some of the most amoral and unsavoury characters ever recorded—even worse than Steve! As a boy I had read about Sweeney Todd, who was running around at about the same time as Burke and Hare and Jack the Ripper, and it scared the piss out of me. Eighteenth-century London was a black pit of evil and Todd, born in 1748, was its most damned offspring. He came from a typical gin-fuelled broken home, and was already in prison at fourteen where he learned, amongst other things – like how to survive – the barber's trade. At nineteen he was released and eventually saved and stole enough to open his own shop. And so the Demon Barber of Fleet Street went to work.

Todd had one barber's chair in the centre of his tiny shop, and another identical chair fitted to the ceiling of his tiny basement. He invented an ingenious pivoting system whereby he could switch chairs simply by pulling a lever. Todd would lock the shop door, slit the throat of his customer, pull the lever, and in a few seconds the victim would be in the basement and the empty chair on the basement ceiling would be waiting on his shop floor for the next punter. Todd would then scarper down to the basement to finish off the victim in a killing frenzy that would put a Great White to shame. Then he robbed them, skinned them and dissected them.

To further the evil, Todd started a relationship with a woman called Lovett who ran a pie shop not far from his barber shop. And when his nefarious basement became full, he transported the remains of his victims via ancient underground tunnels to Lovett, who ground them up and sold them to hungry locals as veal and pork pies. If they were alive today, they would probably be in the real estate business.

I was curious to see Todd's shop but more interested in the location. It was on Fleet Street near Temple Bar, where The Strand and Fleet Street meet, right next door to St Dunstan's Church. Temple Bar was already a London landmark when Todd set up shop. It is the location of a huge Masonic edifice which was originally erected by the Knights Templar, who used St Dunstan's Church. It was this I had come to see.

Some years earlier I had become a Freemason. A great deal of older guys in the oil world practise Free-masonry. Sweeney Todd's shop was a vital connection to the Lodge that I was curious about and keen to discuss at a meeting.

Freemasonry has made the jump out of the Dark Ages, and removed the shackles of its shady misplaced reputation, one it acquired in the years preceding the Cold War. It is a noble and studious organisation, and I have learned a great deal in my time at meetings, not just history or Templar lore but about myself also. There are Masonic Lodges in almost every city in the world, and, should I find myself in a strange place with no contacts, I can call the Lodge and always meet the most interesting characters. Yes, it does have secrets and signs, handshakes and formalities; they are part of a history that goes back centuries and have been kept alive for tradition—something I respect in a world where moral values are traded for anything and everything every day.

On my way back to Brunei after visiting my father, I quickly lost my rose-coloured glasses. My flight stopped briefly in Singapore, where two-thirds of the passengers got off and were replaced by Indian and Pakistani workers. All the construction and hard labour in Brunei is carried out by imported workers. They arrive in droves to dig ditches and haul bricks for ten cents an hour.

One of the two Indian men next to me nudged me and pointed at his embarkation card. It's the card that you have to fill out in flight for the immigration people at the other end: Where have you been? Where are you going? Where do you live? Got any drugs? etc. I was filling out my card when the Indian guy saw his opportunity to get his filled out for him. Going from the grunts and nudges, I guessed he could not read or write English.

I smiled and said, 'Please wait while I finish mine.'

So he slides his card over the top of mine while I'm writing, tapping with a manky (dirty) finger on the tabletop.

'Okay mate, give me your passport.' I filled it out for him . . . Mr Barney Rubble of Number 1 Credibility Street, Toy Town. He was most impressed and babbled something at his mate sitting on my other side, who grinned and slid his card and passport under my nose. Seymore Butts took his papers back without saying thank you, then disappeared towards the back of the plane, returning moments later with half a dozen passports in each hand, that he dumped on my table without looking at me. He settled back into his in-flight movie, blowing his nose on his sleeve for the umpteenth time: he had a bad head cold and blew his nose on everything except a tissue.

I picked them up and dumped them in his lap. 'No way,' I said, shaking my head.

They landed back on my table again.

'Fuck off.' I grabbed them and threw the lot towards the front.

When the passports flew off he screamed, blowing snot down the front of his shirt. Barney took off to retrieve them while Seymore pressed the call button, glaring at me through the snot. The flight attendant arrived, looking flustered, as most of the passengers at the front of her section had just been hit in the head with snotty passports and had pressed their call buttons. Eventually we all calmed down and tried hard to ignore each other.

The meal cart arrived. 'Beef or fish?' the flight attendant asked and faked a smile.

'Beef please,' I said.

'There's no more beef.'

'Then why did you give me a choice?' I asked.

She gave everyone else the fish. Barney and Seymore practically inhaled their mini-meals, both belching loudly afterwards. My meal arrived, but as I was peeling off the metal lid Seymore leaned over me to talk to Barney and sneezed into my food. I had had enough. I stood up on the seat, stepped on Seymore's head on the way out and found a vacant seat at the back where I tried to relax for the landing.

But it occurred to me that the arrival would be a circus if I didn't get out ahead of Barney and Seymore— they were going to be pointing at me in the queue and I was going to be in trouble. I was going to have to get out ahead of them and make a fast exit from the airport.

Luckily I didn't have any check-in luggage, just my small grip bag and, as everyone made a civilised exit from the plane, I broke out into a run. I was first through immigration and customs—nice one—and even had time to savour a backward glance at Barney and Seymore who were brandishing their papers.

There was a strange air in the office the next day. I asked what was going on and was told that one of our people had been diagnosed with testicular cancer following a standard work medical. (He beat it eventually, but endured more than a year of intensive treatment and trauma.) I was due for my medical that week so the news wasn't comforting.

After my checkup the doctor handed me a pamphlet titled 'Testicular Cancer and You'. He said, 'Have a read of that Paul, and familiarise yourself with the self-examination methods.'

'Why? Is there something wrong with my nuts, doc?'

'No, no, it's just so you know.'

The doctor had given the pamphlet to everyone on the crew to read as a precaution, even the Iban guys who can't read too well and just pointed at the pictures and laughed. I stuffed it into my pocket and forgot about it.

A week later I was sitting in the tool pusher's office on the rig, staring at the drill floor through the window. I found the pamphlet on testicular cancer in my pocket, lit a cigarette and started reading. There was a quick guide to checking yourself with cartoon images showing you exactly what to do. The cartoon showed a naked bloke, bent over and scrutinising his ball sac which he was stretching out with his left hand. His right hand was reaching around the back, shining a flashlight flush up against his nuts, the idea being that you illuminate your balls and look for any abnormalities. The pamphlet also had pictures of abnormalities, like those skin cancer booklets that have pictures of melanomas, so you know what to look for.

I looked up at the windowsill and there sat a two-foot long black metal 'Maglite'. It's the Rolls-Royce of flashlights, the one security guards bash you on the head with. It was two in the morning, everyone on night shift was on the drill floor . . . I thought why not, so I dropped my coveralls, grabbed the Maglite, bent over and lit myself up. Everything looked okay, then I noticed my bits were making shadow puppets on the wall in front of me. Distracted by this, and I have to say mildly amused, I didn't hear the tool pusher walk in.

'Boy . . . what in the fuck are you doin' wid my flashlight?'

I dropped it into my undies and in one superfast move pulled up my coveralls and spun around to face him. But my gear only went halfway up because the Maglite hit my crotch, something they don't warn you about in the pamphlet. To his credit the tool pusher listened patiently as I tried to explain and wiped his Maglite on my sleeve. An hour later, the whole drill crew were lined up checking themselves.

Working in Brunei was a pleasure and three years went by too quickly. My time was up, a new project was forecast in the Philippines, and the 'grapevine' had me going. In the meantime I didn't want any of the postings the company offered, as that involved going to areas I had already worked in or places that were too dangerous. So I settled on the best gig up for grabs, in the Middle East.

There was plenty of oil work for me, hopping up and down the Persian Gulf from Oman to the UAE (United Arab Emirates) and on to Saudi Arabia. At the time, the world was in flux; working in oil gives you a finely tuned sense of change, and it was clear that there was something going on. Guys would sit around and debate constantly, and not just the usual small stuff, a coup or tribal conflict, but something much bigger. The last time the walls whispered like that, the Gulf War kicked off within a year. The drilling just goes on regardless. I met a guy who was working on a land rig in Syria at the time. He was happy to work there because everyone got extra danger money. From the drill floor he could see the coalition forces light up the night sky. One day the crew saw an American F-16 fighter jet crash nearby, and a few hours later the downed pilot showed up asking to use the phone.

I spent most of my time in Saudi, and had no problems moving around the region, thanks to my excellent history and having lived in Brunei without any problems for years. Saudi Arabia is ruled by a tribal monarchy and governed by Sharia (Islamic law). It is definitely not just

German
grandfather,
Berlin, 1942.

My father, 1957.

English grandfather, London,
1943.

Erwin teaching his crew, one job at a time.

Waiting for a chopper, Africa, 2003.

Still waiting, China, 2004.

The offshore workhorse Sikorsky SN-61.

Brunei, our home for three years.

Derrick, looking up from the drill floor (stabbing board on left).

Derrick from the outside with drill pipe racked back in stands.

Drill floor during a typhoon.

Close protection, Nigeria, 2003.

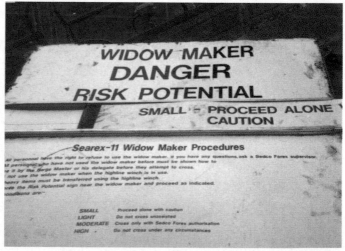

The walkway to the drill floor is aptly named.

Damian discussing confined space anxiety management with Joe, during a visit to Brunei, 1994.

Local visitors, Brunei.

Ah Meng, Singapore.

Live gas well flow test mishap, Philippines, 2001.

Anti-pirate devices,
Nigeria.

another traditional country going through a change. The stakes in Saudi are way higher. It controls a quarter of the world's oil reserves, therefore it can directly affect the supply of oil to the rest of the world. Saudi Arabia is the most prized ally of the United States, whose interest in Middle East politics goes well beneath the surface, and both countries protect their energy fiercely.

This relationship started in 1933 when the Saudi ruler Ibn Saud granted a massive oil exploration contract to the Standard Oil Company. This evolved into Saudi Aramco, the power brokers who have 260 million barrels of oil and 225 trillion cubic feet of natural gas in their back pockets. On top of that Saudi Arabia is the keeper of the Muslim holy cities, Mecca and Medina, and the spiritual home of 1.3 billion Muslims worldwide—that's equal to the population of China. The Saudis have the best-trained, best-equipped military force in the Middle East, and, with the US as their best buddy, who's going to fuck with them? Perhaps themselves; Saudi Arabia is in a cultural maelstrom, where enormous wealth and power meet uncertainty and fear, where tribalism and customs meet mobile phone consumerism. As the oil-rich Saudis wrestle with their small quiet land that we rarely hear about, the Westerners who know the inner politics hold their breath. After the September 11 attacks on the World Trade Center, they shit their pants. Saudi Arabia is the birthplace of Osama bin Laden and fifteen of his hijacker pals.

Men sit in think tanks for months, turning over all the scenarios, so we can maintain our suspension of disbelief that nothing's going to happen again. I wonder what's going to happen in fifty or eighty years when the oil starts to dry up . . .

My time in the Middle East was short. I was in and out, because the walls echoed of war and upcoming turmoil. I opted to get out and go back to Asia, working for myself on a day-rate basis. Freelancing was more lucrative but the work was sporadic and often spontaneous. The phone would ring at ten o'clock at night, and the flight would depart at six the following morning; most jobs started in just that way. As fate would have it, after a few months my old employer phoned and asked if I would 'day-rate' on the Philippines project. I agreed and found myself back with my old crew again.

7

THE HOBBIT HOUSE

1998-99

MANILA IS A MOSH pit with an airport. The city is in perpetual gridlock, and travelling just a few miles can take hours, unless you have a skilled local driver. The trick is to treat the traffic like a bad dog—don't show any fear, because like bees and dogs Manila traffic can smell fear. Drive fast, don't look both ways is a popular technique in Asia; along with if I don't see you, you're not there.

Five of us were in the car getting thrown around all over the place. I'd been up all day in Sydney then got the late-night call, did the high-speed pack, rushed to the airport, flew shitty economy for eight hours, arrived in Singapore, went straight to the job briefing, back to the airport, flew really shitty economy for five hours to

Manila, and was finally sitting in an overcrowded jeep on a congested downtown freeway. The air is around a million parts per million carbon monoxide and we're all smoking. Needless to say I was exhausted, my trick neck had popped out of its socket, but that pain was a welcome distraction from my back pain. I felt like a bag of broken china. I was so tired that potholes, fumes and noise aside, I slept regardless, my head rag-dolling from side to side.

At one point during my deep slumber, our driver hit a bottomless pothole, sending my face into the dashboard.

'Aw fuck!' I sat back, holding my nose, blood running from both nostrils. I angled the rearview mirror to face me and poked at my front teeth. 'Fuck . . . my tooth is chipped . . . fuck.'

As I nursed my smashed face, we progressed up Roxas Boulevard at the rate of your average tectonic plate. The static traffic encourages local kids to wander up and down the lanes selling just about *anything* to motorists. A boy appeared next to me with a wooden box suspended by a rope around his neck. I looked into the box and there, lined up neatly, were a dozen black three-foot long rubber double-ended dildos.

'Back-massager,' the boy said, flipping one around his back and pulling it to and fro as if he was drying his back with a towel.

'No thank you,' I said and smiled.

Peter, who was in the back, grabbed one and tied it

around his head so the two bell-ends jutted out above each eye. '*We mean no harm to your planet,*' he proclaimed with a straight face to the people in cars around us, who looked on in mute fascination.

Ambu thought that was great and also tied one around his head. Then they proceeded to bash each other over the head with them all the way to the hotel, finally stuffing them into their offshore bags as we pulled up in the hotel entrance.

The job was high profile and despite the logistics of travel we were well looked after. The hotel was five star, with gold taps, butler service, the works—which made a nice change from the whorehouses we usually ended up in. After checking in, I watched a porter cart off Ambu's bag across the expansive marble lobby with one foot of black-rubber penis wobbling in time to his efficient walk.

Peter and I were in the elevator on our way up to our rooms when a Japanese family got in. Peter is truly an animal. He takes tremendous pleasure in embarrassing me, and we had just departed the second floor when he rocked over on one leg and let go with the biggest fart I have ever experienced. And it was an experience because not only was there the expected olfactory-audible result, but the looks on the faces of that poor family are something I'll never forget. Mum and Dad went bright red, exchanging inner screams, as Peter finished off. My ears popped. One of the two children screamed and buried his head in his mother's skirt, the other child started crying. I was as red as the parents, and mouthed the words best fitting Peter's abilities: 'You bastard.'

He grinned and looked at the crying child and said, 'Hey . . . there's a monster in here. I can smell him.'

We all held our breaths to the top floor, but the damage had been done and those children would never be the same again.

The next morning the rig phoned: all choppers were cancelled because of bad weather. 'Hurry up and wait.'

I spent the day with Tommy, the driver. He's a speed addict and made no attempts to be discreet about it. He did, however, get me from A to B faster than any bus or taxi could.

'I drive car like the wind yes,' he said through clenched teeth.

Tommy also talked almost as fast as he drove the car. I heard all about his battered past and underprivileged siblings doing the hard yards in the provinces. He explained in high-speed pidgin English how he had been too high to sleep the night before so decided to clean his flat at two in the morning but got caught up in finding out just exactly how much toilet paper you can suck off the roll with a new vacuum cleaner.

Tommy was good at his job, but he was also armed. I was more than a little conscious of him being on drugs with a gun in his pants. Then again, almost everyone in Manila is armed, and some bars even have a little booth like a cloakroom where you have to 'check your weapon' before entering.

I asked him to show me the real Manila: he did. At one point we drove into a ghetto back street lined with garages. It was like stumbling onto the set of *Night of the Living Dead*. All the junkies staggered out of the darkness with bloody bandages on their arms and legs. Tommy explained that heroin is cheaper and easier to get than a needle, so in desperation they slash into an open vein and rub the dope into the wound.

He also took me to his local cockfight arena. Cock-fighting, like not getting run over, is a national pastime. The birds are bred to fight, so imagine a vicious chicken. The pit is lined with clear plastic that is splattered with blood, because the birds fight to the death. Small curved blades, each about 3 inches long, are tied to the bird's legs so the fights are short. There was lots and lots of betting going on. It took some time to get my bet down as everyone was yelling at everyone in Tagalog (the Filipino language) and I had to point a lot and do a pantomime of 'Rocky'. But I ended up winning 500 pesos on a particularly violent chicken who, with his huge puffed-out plumage, looked a bit like Tina Turner.

Tommy dropped me off at the hotel and said he expected us to be standing by for a few more days as the weather forecast was really bad. Filipinos tend to be a little blasé about bad weather. Of the seven thousand islands that make up their beautiful archipelago, not one escapes the average thirty-three typhoons a year. On top of that, the Philippines has seventeen active volcanoes and regularly experiences earthquakes that would send your hardened Californian tremor veteran under the nearest door frame.

The next day was miserable. Monsoon rain streaked down through the smog, turning Roxas Boulevard into a river; no adventures today. Peter came by all excited about a bar he had discovered the night before. He really wanted me to go with him that night, but I told him that I had seen it all, every weird sick gimmick that any Asian bar had to offer. Peter assured me I had never seen anything like this, but he wouldn't tell me what it was. Coming from a man who still does fart tricks at forty-five years of age, it didn't capture my imagination.

We watched an in-house movie and ordered room service for lunch. When the room service guy arrived at the door, he looked a bit pale. I asked him if he was okay, and he said there was a hurricane going on outside, then opened up the thick floor-to-ceiling drapes to reveal a horrific storm.

'I'm worried about my family,' the room service guy said.

'Fuckin' hell,' said Peter and opened the balcony door, but the wind snatched it out of his hand and slammed it back against the wall.

Wind and rain blasted in, almost knocking over the little room service guy. All three of us pulled the door shut, and just as I closed the latch we watched a palm tree sail past the balcony. Our hotel had double-glazed doors and thick walls, one of the few buildings in the area that wouldn't fall down if you pissed against it. But half of Manila is a shanty town, and a lot of people would be left homeless.

I went back out on the balcony. The wind was so strong it was creating a dead air vacuum that wanted to suck me out into the vortex. I gripped the handrail and looked over. The storm sent bits of corrugated tin and debris in a procession past the hotel. The palm trees were bent over almost touching the ground, and the streets were now awash with stalled cars and blown-off rooftops.

That night Peter returned to my room pleading with me to go with him to the bar he was obsessed with. The weather had died down but still no choppers for at least the next twenty-four hours.

'All the boys are coming except you. C'mon, we'll get loaded . . . have a good time.'

I gave in. Tommy arrived to drive us through the rain to Manila's red light district, Makati. Peter told Tommy to drop us on the main strip, and then led us down a side street to an old timber and brick building. Its neon sign flickered in big pink broken letters: 'The Hobbit House'.

We went in through two saloon-style doors that opened into a large room with a big bar at one end. It took me a moment to take it all in; we just stood there. The place was entirely staffed and run by dwarfs and midgets, it looked like every tiny person in Asia had a job there. Peter was right. I hadn't seen anything like it.

'Hey, where's Ambu?' I asked.

'He's still outside . . . Fuckin' great place, huh?' Peter was right at home.

Ambu was talking to Tommy who had parked the car, and they were still standing on the main street. I went to get them. As we walked up the side street Ambu froze. He peered at the entrance where two midgets were now standing. 'How far away are they?' he asked.

At the bar stood a driller, Paul, who I hadn't seen in a few years. He was a big man in his forties, and he roared when he saw me then picked me up and hugged me. Paul was drunk, and more than a little upset over his recent divorce, as I learnt over the next hour.

A midget prostitute in heels and a boob tube was walking on the bar. She sauntered up to Paul and me, stopped in front of us, leant a hand on my shoulder and at eye level said, 'You buy me drink.'

'Err . . . some other time.' I didn't know what to say.

Paul was laughing. 'You think that's funny, check him out brother.' Paul was pointing at Ambu, who, having recovered from his initial shock, was chasing a midget wearing black velcro coveralls, a neck brace and a kid's crash helmet.

'You pick him up 'n throw him against that wall,'

Ambu said with excitement as he ran past towards one corner of the bar which was padded and covered with velcro from floor to ceiling.

We did shots of Tequila and Paul was in the middle of another bitter divorce story when the little guy in the crash helmet ran past. Quick as a flash Paul grabbed him by the ankle and, a beer in one hand and the upside-down midget dangling in the other, continued his story.

'Anyway, I said, "You take the kids bitch and I'm not givin' you shit".'

I interrupted him, 'Mate,' and nodded at the poor midget; all the blood had rushed to his face.

'Oh yeah,' said Paul and, in a kind of drunk hammer-throw manoeuvre, spun around and sent the midget flying end over end into the wrong wall.

Everyone booed and threw food and drinks at us.

The night degenerated further into full-blown drunkenness. I managed to escape with Tommy, who was as drunk as me, and woke up outside my hotel room door, a porter shaking my shoulder.

'Sir . . . sir.'

I looked at my watch, it was two in the morning. 'My card key won't work.'

He bent down and took it from me. 'You're in the wrong hotel, sir.'

They were very understanding and organised a taxi to take me to the right hotel where I promptly threw up.

It took two days for me to feel normal again. The weather had cleared up and our chopper was set for an early departure. Our flight to the rig was going to take two hours, with a refuelling stop on one of the bigger islands called Busuanga.

Sparsely populated, Busuanga is a lush green paradise, spectacular from the air. While the chopper refuelled we had just enough time to wander to a small shack where a woman made the best noodles I had ever tasted. Eating them was done by holding the bowl under your nose and pretending to use chopsticks while actually slurping down your food with lots of loud noises. It's considered bad manners to eat Western style, i.e. with cutlery, but belching, farting, picking your nose/teeth/bum, smoking, spitting, shouting and kicking the pig that's under the table are all perfectly fine.

At the time, the Philippines wasn't the safest place to be; there was an extremist group creating havoc in the southern islands. The group operating where we were was called the 'Abu Sayyaf' and specialised in K&R (kidnap for ransom), usually of tourists who would finish up beheaded on TV.

Oil workers have to have very complex insurance policies that cover everything from acts of God to getting snatched by extremists. K&R happens a lot in the oilfields, especially in South America, Africa and parts of Asia. The extremists know they will get paid the insurance money and so prefer to grab oil people, whose insurance costs go up as more and more get kidnapped.

This cost eventually finds its way to the oil companies, who countermeasure by hiring third-party security specialists. They are tasked with providing close protection and defend the rig and its personnel. Depending on who you're working for and where, these people vary from thugs with guns, to mercenaries, to ultra-professional ex-elite forces personnel who tend to be of medium build and height, are quiet and organised, and to all intents and purposes look like my accountant. Except they can perform complex mental calculations while constructing a shape charge to breach a hotel fire door that's been welded shut, and drop someone at 500 yards with open sights or a toothpick at three feet. And, of course, they can maintain control of half a dozen shit-scared rig pigs.

Despite all this, the rig was a pleasure to work on and it was great to be back with the crew.

One of the drillers was a Frenchman who thought it was great fun to try and aggravate me by attacking the monarchy.

He would bound up and yell across the drill floor in his thick French accent, '*Hey Paul . . . Fuck your queen . . . ha ha ha ha ha ha.*'

Or he would page me, and the speakers located in every room on board would chime: 'Bing . . . Bong . . . Paul, pick up . . . Line one please.'

And I would dutifully jump to the nearest phone, 'Paul speaking, hello.'

'*Fuck your queen . . . ha ha ha ha ha ha.*'

Then my new French friend, who I now called 'Frog One', progressed from verbal attacks to practical jokes, starting with the old styrofoam-cup-dipped-in-a-grease-bucket-and-carefully-stuck-on-your-hard-hat-when-you're-not-looking trick. I walked around with half a dozen cups on my hat, the boys grinning at me. 'What?'

So I retaliated with the Fork in the Redwing. This is a good one . . .

First you take a fork and place it inside a safety boot, on its back so the prongs are facing up. Redwing boots are preferred as they are big, heavy and fairly high with two looped leather straps that you hook your fingers through to pull them on. This requires some effort: you have to really push until your foot reaches a point of no return and finally it slips down into the heel. If there is a fork laying there, you end up with the fork's prongs lodged behind your big toe. If you try to pull off the boot, your toe gets stabbed. The only way out is to cut through the steel toecap and pull out the fork through the sawn-off toe, wrecking the boots. It takes ages to saw off the toe, especially because you have to hobble all over the place looking for the hacksaw that's been hidden by the bastard who put the fork in your boot.

'Fuck you,' the French driller yelled down the phone.

'Fuck your president,' I replied calmly.

He got me back a week later. We had to go on stand-by in town and wait while the rig finished drilling out the next well section. So we all got in the personnel basket, which is a big metal basket for transferring people from the rig to a vessel such as a boat. The crane operator picked us up, swung the boom over the side and instead of setting us down on the deck of the crew boat dunked us into the sea, waist deep, for ten minutes. I looked up, dumbfounded, and then saw the Frenchman giving me the finger from the crane 200 feet above us.

I could just make out his mouth, 'Fuck you'. He was ecstatic.

The crew, on the other hand, was really pissed off. We were soaked, as was all our gear, and the ride to town was going to take twelve hours.

A week later we arrived back on the rig and I was prepared for an awesome payback. While in town I picked up two big bottles of food dye. When Frog One finished his shift, I found his new boots and emptied the food dye into them. After a twelve-hour shift in the tropics, your feet are soaked, as is the leather inside your boots, and they are often still wet when you start your next shift. He came out in the morning, checked his boots . . . he knew I had arrived the night before, but there was nothing unusual and so he began work.

Twelve hours later I heard a scream by the boot rack. He was going to have one green foot and one blue foot for weeks.

'Paul, pick up . . . you bastard . . . English pig.' He was on the PA so I answered.

'Good evening Frog One.'

'Fuck you . . . I get you back you bastard.'

I thought I got away clean, but a week later I was on another job in another part of the world when one of the roughnecks ran up to me.

'What happened? Sit down,' he said, then took off my hard hat and burst out laughing.

'What the fuck are you doing?' I protested.

Frog One had soaked the black foam inside the sweat band of my hard hat with red food dye. As I sweated it ran down my face and looked like I was bleeding. And I had a red band around my bald head for weeks.

The Philippines offered up a lot of surprises towards the end of my two years there. The Abu Sayyaf had become a real problem for expat personnel. We continued staying at the same hotel, the staff were very nice and efficient, and would try to get you back if you got kidnapped.

However, we could no longer stray too far from the hotel as the terrorists had begun a hardline bombing campaign: detonating devices in shopping malls, outside churches, in the international airport, in rubbish bins in crowded streets. And there was the regular decapitation of some kidnapped tourist splashed over the papers and TV.

Standing by in the hotel one day, I ran into Dangerous Dave, an American ex-army man; if Dave was any more laid back, he'd be dead. After four consecutive tours in Vietnam, Dave just doesn't sweat the small stuff. I think that's the way he's been since the early 1970s; he can recall the most horrific moment and still smile afterwards. I liked Dave; he never acted dangerous, but wouldn't tell me how he got his nickname. We would chat for hours in the hotel bar; Peter would join us but only until Tommy was ready to drive him to the Hobbit House.

After one particularly enjoyable session with Dave, I wandered up to my room and fell into a deep sleep, only to be woken by a nauseous feeling. I hadn't had that much to drink, but I felt like I might throw up, so I sat up in bed, turned on the bedside light and rubbed my head. The curtains were moving, but it didn't register. I got up and felt dizzy. The curtains were definitely swaying from side to side. I stepped out onto the balcony, hundreds of car alarms shrieked below and I realised that the whole building was swaying. It was an earthquake. Shit.

I tried to remember what to do. I grabbed my wallet and passport, threw on a hotel bathrobe and took off down the fire stairs. Passing the lift door I could hear people crying out from the long shaft below. Taking three stairs in one step I made it from the top floor to the lobby in record time, blasting through the side door into the crowded lobby. The hotel manager was reassuring a throng of guests that the hotel was built on those rubber shock absorber things that let the building move rather than fall down.

'They have been used in America for some years, and I can assure all of you that our hotel is built to withstand . . .'

Crack.

He stopped as the massive chandelier above us began to break free of the roof.

Everyone scattered. I jumped over the reception desk, hitting the floor at the same time as the chandelier. But there wasn't a huge explosion of breaking glass as everyone expected, more of a thump. The whole thing was made of some kind of safety glass and it basically just bounced about for a while, a bit of an anticlimax really.

I found Dave in the lobby examining the shatter-proof chandelier.

'Pretty cool,' he said and smiled.

I bummed a smoke from him and suggested we have a drink. 'I'm not going back up to my room just yet.' The quake was over, but the whole aftershock thing was playing on my mind. 'Let's sit out the front,' I said, 'Then if there's another one we can just grab our drinks and run across the street and watch the hotel bounce.'

We had just sat down when the petroleum engineer from the oil company came bounding up to our table. In her late forties—I can't remember her name, I must have blanked it out—she was a nightmare, the only aggressive Vietnamese woman I ever met. She worked in the office south of Manila and had been staying in the hotel waiting for her flight home in the morning. Dave kept his disdain for the woman to himself; always a gentleman, he just smiled and sipped his whiskey. She was in a state over the quake but suddenly digressed into giving Dave a hard time about his company's equipment failure on the rig, something Dave was totally uninvolved in.

'Hey,' I said. 'He's been standing by in town for the last week, talk to his back-to-back on the rig in the morning.' ('Back-to-back' is the man on the rig who does your job when you're not there.)

She put her hands on her hips and glared at me, not sure what to say.

'Okay, bye-bye then,' I said and glared back.

There was a short burst of high-speed Vietnamese then she spun on her heel and stormed off.

'What did she say?' I asked Dave.

'You don't want to know,' he said, watching her walk off. Then he added, with a smile, 'Could have sworn I shot her in the war.'

My jaw dropped.

'Just kidding,' he said.

Dave told me stories about the war. He had seen a lot and I guess his number just wasn't supposed to be up then. He talked about Saigon, the good times he had. I asked if he ever went back after the war.

'Sure,' he said then explained how it just made him feel old as the few former girlfriends he managed to find had all turned forty and had bad backs.

Dave was not a grumbling veteran; if anything, I had to really push him to talk about the past. He always painted an intriguing picture of Vietnam—his backbone stiffening at the memory of long-gone combat—and I hoped to go there and see it for myself one day. Once again synchronicity put me there on my next campaign.

I took a break in Sydney, riding my motorcycle all over the place, exploring. Stopping for a drink one night I watched a stunning young woman walk into the bar and in one fluid movement peel off her jacket and hurl it on top of a giant fridge.

Her name was Pia, and she was as bright and funny as she was beautiful. I spent the rest of my time in Sydney with her. I was so caught up I missed my departure flight back to work for the first time in ten years. She was all I could think about for the next month. I was in Singapore, preparing the tools for Vietnam, mundane stuff, and I phoned her every night, feeling unable to wait a month to get back to Sydney. When I did get back, the time passed way too fast.

Pia's family were superbly hospitable. We would spend weekends together and I was completely at home in their company. For the first time it hurt me to have to go offshore, and I realised how hard it was to have a family and work in the oil industry.

That was a bad moment. If this was true love then I understood the agony but not how to deal with it.

8

WHERE YOU FROM?

1999-2000

I ARRIVED IN VIETNAM without incident, even my flight was comfortable. However, I then had to hang around the airport for two hours, waiting for a colleague to arrive from the United States. John was a Texan, he sounded nice on the phone, it was his first trip overseas, and I had been assigned to babysit him until he got to the rig.

In the past young men in the oil industry were pulled like magnets to the neon lights and semi-naked teenage prostitutes of Vietnam, often on the way from the airport to the hotel, only to end up in trouble, handi-vacced of all documentation, money, jewellery, vital organs, and they hadn't even seen the rig yet. This is done in all manner of slippery ways, my favourite being the 'Titty Mickey': this involves some kind of powerful God-knows-what sedative applied to a nipple, said nipple is then thrust into the face of Western John Doe who thinks 'Super', and two minutes later he's face down.

I found a comfy place to sit with a good view of the arrival hall. I always travel in jeans and an old T-shirt, trying not to look like anything but a backpacker fast approaching his use-by date. I couldn't have missed John: he was six feet plus, well built, in his late twenties, wearing a Stetson and the obligatory hubcap-sized belt buckle; he had the confident cowboy-boot stride of a man who knows how to rope a barmaid; his Halliburton alloy briefcase was covered in oil company stickers, and his Rolex and puzzle ring twinkled 'MUG ME' across the airport.

As I approached he looked at me the way you would if a stranger asked for loose change. I talked him into waiting inside the airport while I got us a taxi—one look at Big Tex and the price would quadruple.

We dumped our bags in the boot, I told John to keep his briefcase with him, and if possible to sit on it. The taxi was an old Chevy with no airconditioning. As we got in he wound down his window and laid his ten-thousand-dollar arm on the sill. I tried to explain why that wasn't a good idea, and the taxi driver also suggested he wind up the window, but John just spat his wad of chewing tobacco out the window, gave me an ambiguous grin and said, 'Aw hell, no one's gonna fuck with me, old buddy.'

I smiled back and locked my door. Ho Chi Minh City is full, to the smog-filled tune of six million people. We coursed down tree-lined Phan Dinh Phung Boulevard, past French villas, with their wrought-iron gates green with age, then went deeper into the heart of 'Old Saigon', which is local jargon for the city's increasing decline into drug trading and prostitution. To get to our hotel we had to drive through a seedy part of town, navigating through a few million people on scooters, and before long the wide boulevards gave way to narrow backstreets.

Amputees scuttled up to our car, begging for change. Street kids pounded on my window. The driver yelled at them, 'Bui doi (dust of life), no good.'

John was throwing US coins out the window, much to the dismay of the kids who threw them back as no one trades in coins, only paper money. In this grim sanctuary for panhandlers John got nailed. Two young men on a scooter roared up, the pillion jumped off and held a machete under John's throat, while a third came from nowhere and stripped him of everything—all in a few seconds. There was nothing any of us could do, with John sitting there, the blade rammed against his throat pushing him up until his cowboy hat was pushed down over his eyes. In a plume of blue exhaust smoke they were gone, and so were John's passport and offshore pass. Oh well, I thought, that put me on the job alone.

The driver turned around, shaking his head at John, and said, 'We go embassy now.'

'Hotel first, then you take him to the embassy,' I instructed.

John was in shock, he wanted to get out of the car and chase them, but they had long gone. 'You're lucky they didn't take your arm off,' I said.

The next day I left John at the US Embassy and continued on to the industrial area of Bien Hoa, once a massive US air base during the war, and from there to Vung Tau, a crowded town where all the supply boats gather to feed the rigs. I found my hotel near the harbour and phoned the rig, explaining that John wasn't going on the job and there was no one else obtainable at short notice. They understood, so I stood by in the hotel waiting for the rest of the crew to arrive later that day.

The crew members were flying in from all over the place, arriving one by one at the hotel. Erwin and Ambu, amongst others, were involved, and I was stoked to be with them again. I left a message for the crew to meet me in a bar nearby, and there we stayed for the next three days, waiting for the rig to finish delayed operations.

On the second day Damian came down to the bar and sheepishly stashed a pink shopping bag under the table. Another one of the guys grabbed it and pulled out a skipping rope.

'I'm fed up with jogging around the heli-deck,' explained Damian.

'You can't skip,' said Erwin.

'I can . . . I can even do the crossover thing.'

'Go on then, Leon Spinks.' I handed him the rope.

It wasn't any old skipping rope mind you, this one had an elasticised rope and weighted handles. Damian stood in the middle of the rundown bar, directly underneath a huge ceiling fan. I was about to say something but Erwin winked at me, and I thought what the hell, it could be funny, so shut my mouth.

Damian swung the rope in an 'I could have done this for a living' confident way, only to wrap it over the centre of the fan which then snatched it from his hands, whipping the rope at head height around the bar. Damian hit the deck, hands over his head. Everyone stopped laughing and took cover when the momentum of the fan stretched out the elastic rope, sending it swinging in our direction. The helicoptering handles began smashing glasses on the bar as the fan started to come out of the roof. It eventually fell, landing on top of Damian, who just got up and walked out, but refused to speak to us for days.

The afternoon of the third day finally found us walking over the grassy airstrip towards the chopper on our way to the rig. The small airport was still riddled with thirty-year-old bullet holes, and everywhere you looked brass casings littered the ground. From the air, however, we got a much better idea of just how much ordnance was dropped on South Vietnam during the war.

The huge craters left from the bombing had never been filled in, just overcome. Rickety bridges had been constructed over any that landed on roads, the rest were now ponds.

'Fuck . . . they shelled the shit out of this place,' Damian said with his camera pressed up against the chopper's perspex window.

The rig was nasty, with cramped dirty rooms and old smelly toilets. I had a shower after we arrived, that is to say I stood under a nozzle that jutted out of the wall dribbling water on my head. The TV room wasn't much better; you never know what you're going to find when you walk into a rig TV room, CNN, BBC, Discovery Channel, naked prom queens cavorting in jello. Then it was off to the galley for something to eat. The crew had renamed it 'Chucks', the food was bad, even now when I burp I can taste it. I shuffled along in the queue, my aluminium tray in hand, playing 'Spot the Con' with Erwin.

Spot the Con passes the time while you're waiting for your gruel. There are a reasonable number of ex-convicts working in the oil industry. You can usually pick them out by their rough jailhouse tattoos, and the way they protect their food: the ex-con is the guy who looks at everything except what he's eating without making eye contact.

I got to the service window and found a three-hundred pound, sweaty, bald Chinese guy glaring at me over the bain-marie. Had he been wearing a black one-piece suit and a bowler hat, he would have looked exactly like 'Odd Job' from the Bond movie *Goldfinger*. I asked him to identify the various grey reconstituted meat food products bubbling in front of me. He began rattling off the menu as Erwin leaned in and started doing his best Sean Connery, but Odd Job had obviously heard the joke about who he looks like before and let fly with some high-speed abusive spittle-ridden Cantonese.

Dinner was foul, it had all the texture of a boiled shoe and even less flavour. I went to bed miserable, pulling back my threadbare sheets to reveal a cockroach that resembled a bronzed soap bar with legs. I thought, I will dream and escape to wide rolling hills and quiet walks with Pia, wake up happy and refreshed if Ambu doesn't snore, rig up and start the job. The sooner it's over the sooner I'm home.

The Vietnamese crew was great to work with, the job was going well and we would have all the pipe run in the hole and be back on the beach in two days—nice one. I started my second shift at night; a nasty storm was blasting the rig and the seas had really picked up. It was raining hard and the rig was moving around a lot. This makes the derrickman's job even harder; Ambu was up there doing his best. The driller, on the other hand, didn't see it that way. An American with no patience and a short temper, he got fed up around three in the morning, told the assistant driller to take over, and stepped out from behind the brake.

The brake is a large metal lever jutting out of the floor at the driller's feet. It can be locked down into place via a chain and, as the name suggests, the brake stops the movement of the top drive, and in turn the up or down movement of the pipe which is being made up (screwed together) on the drill floor.

The assistant driller took hold of the brake while the driller came over and told all the Vietnamese roughnecks to step back. The local roughnecks are small people and we were running a 13⅝-inch casing, a big heavy pipe.

With the movement of the rig and the small people trying to line up the free joint hanging in the derrick it was all too heavy with too much movement. The driller got hold of the loose joint of pipe, swinging about in the derrick in a manly kind of vertical head lock, and started heaving and pushing it over the static pipe sticking up in front of him. Finally the two pipes looked lined up, separated by only six inches and the driller's beer gut hanging over the inside rim of the static pipe. The driller had his back to the assistant driller, who let the loose pipe down and they lined up, slamming together with a loud metallic clang.

The driller staggered back but at first none of us realised what had happened. His arms were doubled over his belly, then he straightened up, dropping the contents of his abdomen on the floor at his feet. He had disembowelled himself, and was dead within seconds.

Because of the weather it was out of the question to get a chopper and send his body home, so the medic taped him up in bin liners and put him in Odd Job's cool room. The chopper we were expecting was a Bell-type 212, the same model used in the war, and luckily someone realised the driller, a really tall man, well over six feet, was too big to lie on the floor as his feet would stick out the doors. The seats were welded to the chopper so moving them was impossible too. Then we hit on the solution: we moved him into Odd Job's freezer, taped him to a plastic chair and froze him in the sitting position. When the chopper finally arrived for him, the roughnecks gathered around his rigid body strapped into the seat and had a group photo while he began slowly defrosting.

Back in Vung Tau after the job the crew amused themselves with a shocking TV show that can best be described as 'Cambodian Idol', featuring poor lonely people with no legs singing on stage—all the contestants had some kind of prosthetic appendage. Erwin and I opted to go and visit the Cu Chi Tunnels, 70 miles north in the Tay Ninh Province. There is a visitor centre there where you can see, according to the Vietnamese information posters, why America failed to bring US-style democracy to Vietnam. For five dollars you get a guided tour of the famous tunnels, a subterranean maze of Vietcong guerrilla caves, and for a dollar a round you can shoot a surplus Soviet-made AK-47 or an old M-16 at a paper water buffalo. If you're a good shot you win a Vietcong hat.

At least some money is coming into the country by turning the remains of war into a tourist attraction. More than half of Vietnam's current population was born after 1975 so there is little interest in the conflict from locals. We had beers in a bar called 'Apocalypse Now'. Erwin told me how he got a role as an extra when Michael Cimino shot *The Deer Hunter* in 1978; later at home I got the digitally remastered DVD and sure enough there was Erwin. We kicked on to a few more bars, having a good time, until a group of drunk American businessmen called us 'oil trash' and shoved Erwin. We were not dressed like oil trash, we didn't draw attention to ourselves, or look for trouble. He ignored them until an olive bounced off his head. I had never seen Erwin really go off before: he waded through them like a Jedi master, and we got arrested.

The Vietnamese police are not to be fucked with. They cuffed us both and took us to the station, where we sat and waited, cuffed again to a huge old wooden bench. Metal loops jutted out of the middle, creating an armrest as well as a solid place to attach the handcuffs. Both your wrists were cuffed with the metal loop running through the chain, so you had to sit with your hands to one side. An hour later the bench was almost full. I had a small local guy sitting next to me on the right, Erwin was next to him, and he was next to a prostitute and two very quiet Europeans. The only vacant seat was on my left. Three police officers crashed through the metal doors, struggling to subdue a massive American. He looked like a biker, with a long greying beard, shaved head and tattoos up and down both of his massive arms. He roared obscenities at the police, who flayed about at the end of his hulk, until finally he got tired and they cuffed him to the seat next to me, but only by one hand.

I tried hard not to make eye contact or even a sound but after a moment the biker grabbed my shoulder in his paw and twisted me around. 'DID YOU FUCK MY WIFE?' he spat out through brown teeth, veins bulging on his melon head.

'No . . . no I didn't.'

He let me go, leaned forwards and looked down the bench. 'DID ANY OF YOU MOTHERFUCKERS FUCK MY WIFE?'

Everyone froze. Then Erwin leaned forwards, looked him in the eyes and said, 'What does she look like?'

The biker was on top of me in a second, trying to get to Erwin with his free arm. The whole bench came down, and soon the police who brought the biker came in.

'He started it,' I said and nodded towards the biker thrashing about at the end. So they stun-gunned him until he pissed his pants, unconscious.

My flight back to Sydney was just as smooth as the trip in, unusual for me. Pia was going to visit relatives in rural New South Wales and invited me to go with her. We spent two weeks on a fantastic farm located in a picturesque valley; they had motorbikes I could take off on, and rifles to shoot tin cans with. Pia spent the days painting and in the evenings we would go for long walks.

One day a bull arrived in a big truck. He was a prize bull and everyone hoped he would father lots more. After a few days, however, it became apparent that the prize bull was gay, opting to ignore all the best-looking cows and bash down a fence made of telegraph poles to try and hump another bull, who ran away, seeking refuge in the paddock where the farmhouse was located. We woke early the next morning, the whole house rocking from side to side. Not another fucking earthquake! I jumped out of bed and ran out the front, grabbing a banana on the way so I could have breakfast and watch the house bounce. Standing on the front porch was the 'straight' bull; he was having a scratch against the wall and moving the whole building. He really was a big animal.

'Where's your boyfriend?' I said in his ear. He wandered off unimpressed.

9

BARREL FEVER

2000–01

MY PHONE RANG IN the car on the way back to Sydney. There was a job in Papua New Guinea on a land rig in the highlands and my flight departed the following morning. Again it was a cakewalk. I'm on a roll, I thought. But not for long.

The rig was right up in the mountains, deep in the jungle. Locals wandered up every day in grass skirts, carrying muskets, their faces painted. Now and again they would take a pot shot at the rig. The location was still being set up, the area having been cleared using high explosives, otherwise known as 'instant wood chipping'. In every ancient tree, wise and proud, were generations of evolution buried deep, which is vaporised 'cause it's fast and cheap. Even though they plant another tree, somewhere else, to make up for the one they destroyed, I feel a twinge of guilt, because essentially I'm a cat-loving pacifist who ought to care deeply about the environment. On the other hand, I represent people who would squeeze schoolchildren to death if they thought some oil would come out.

To summarise my political opinions about oil, greed and the environment, both then and now:

- I firmly believe that when politicians aren't kissing babies they're stealing their lollipops.
- There is no oil in schoolchildren.
- Everyone in oil is a lying weasel . . . except me.

Lots of guys in the oil industry have bucked against the system until the system has hardened them into balls of frustration and that's all they've got left. The environmentalists and the oil companies both have their propaganda machines vilifying each other, but inevitably most of the time we fall back reassured that the entire planet is nothing but two feet of top soil surrounding a huge ball of oil. In PNG you can exploit the locals; you can do whatever the hell you want with government backing and your own armed security. Sometimes the oil companies do go into overkill . . . fear of the locals overwhelming the rig has the security personnel asking if they can mine the perimeter, and after compromising decide instead to use two hundred bear traps from

Canada. (At the time I wondered if they got the traps from the same company that supplies the clubs for bashing baby seals.) But mostly, we do need the heavy security, as proven not so long ago in Banda Aceh, at the top of Sumatra's long finger, where the local people tore a rig to pieces.

The first time an attack happened I thought, all these guys with guns is a bit much isn't it? Like using mercenaries to discipline naughty schoolchildren or hiring Jamie Oliver to help Pol Pot eat people in Cambodia. But I was thankful for their presence on more than one occasion. Most prevalent in my mind is a drill floor in Columbia. At the time Bill Clinton was 'fighting the drug war' but using the DEA choppers to ferry men and equipment to and from the rig. So apart from getting its dope from Columbia, America gets quite a bit of oil there as well.

The rig was a bit close to the other major cash crop in the region. As a result some unlucky third-generation farmer got his straw-thatch hut, goat, wife and crop napalmed. Naturally he's angry and finds the rig, levels his weapon at it and has a go. The alarm went off; a horn and light flash which meant run fast into a safe room, lock the door and wait for the security staff to give you the all-clear. Just before the alarm sounded a thirty-foot mushroom cloud exploded in the jungle near the rig. It turned out another murderous farmer/local gunman had fired a Rocket-Propelled Grenade at the rig and missed, the RPG passing directly between the cross members of the derrick. The security staff took a back bearing and hunted the man down. He was in two body bags in the cool room within the hour. I asked why two bags and was told that they had used hot loaded strung buckshot. This is a solid projectile that's been cut in two, hollowed out in the centre, with six inches of wire coiled inside and the two heavy ends spinning around the wire. It cut trees down; from the top of the rig you could see a cleared tube through the jungle.

On the highlands rig we managed to avoid life-threatening locals; same politics, though, but this time PNG got the last laugh. We had just finished powering up our tools and checking everything when one of the rouseabouts came up carrying a case of bottled water. We all grabbed one each and guzzled it down. Anywhere in this part of the world it's second nature to be wary of the drinking water: I have even been caught out with bottled water when I put ice in it and fell sick because of the ice.

After only half an hour or so I felt ill. One by one the crew began to disappear, doing that hurried clenched run of a man about to have a rapid bowel movement. My turn came, I took off across the deck, there was no warning, everyone was in the toilet blaming the bottled water. I went to the medic who examined the case of water still sitting out on the deck. He realised what none of us did, the bottles had been filled with God-knows-what and re-sealed. Small beads of superglue had been applied to the ringseal that attaches to the cap, so we got the loud crack as we unscrewed the caps, giving us the okay to gulp down a litre each.

The medic looked worried. He explained that the necessary medication to deal with this was not on the rig yet so he was going to get some immediately from the supply base in Port Moresby by chopper. He was still on the phone organising this when our condition progressed from regular diarrhoea to red hot violent uncontrollable I-can't-leave-the-toilet diarrhoea.

The medic ditched Plan A and organised an emergency chopper to get us to a hospital, as he feared we had all contracted dysentery. In a situation like this you're given the choice of going to a local hospital and having the wrong procedure performed by someone who can't speak English, or flying to the nearest Western hospital, first class if necessary, and having your bits sewn back on or whatever you need, assured in the knowledge that you probably won't die. I chose the get-out-of-PNG option, thinking stupidly that I could clench all the way to Singapore. Three of the others chose possible death by bowel movement in a local hospital.

We boarded the chopper with our IV drips inserted and towels shoved down the backs of our coveralls. Transferring directly to the first flight to Singapore, I was in business class next to a large happy-faced German man whose expression dissolved into horror when he saw the IV. We took off . . . so far so good. But about halfway through the take-off climb my backside let go. I yelped in complete terror—I'd just lost my arse on a commercial airliner . . . oh my God.

Two scalding squirts of piping-hot poo shot down both trouser legs. I feverishly pulled at my seatbelt and, grabbing the IV in my fist, hurtled down the fuselage towards the virtual heaven of a business-class toilet. The flight attendant looked sympathetic as I shot past and thankful, I'm sure, that my trousers were tucked into my boots. Slamming the door I spun around, gripped the IV bag in my teeth, pulled down my coveralls, disconnecting the IV bag from the needle, and sat down just in time for round two of the most embarrassing experience of my entire life.

I think that over the next hour I must have shat my own bodyweight, and then the projectile vomiting started. I don't know if you've ever been violently sick in an aircraft basin, but in case you haven't, don't, because it flies straight back out and all over you.

By this time I didn't know if I should sit or stand, eventually opting for the more comfortable vomit-on-your-own-genitals-position. I refused to come out, naturally, no matter how much my colleagues begged. 'Piss off, find your own.' They had to tag-team it, using the other toilet opposite me all the way.

When we did finally get to Singapore, the aircraft pulled up short of the terminal, a staircase on wheels arrived at the back door and an ambulance was standing by. We debussed to the appalled looks of our fellow passengers; a few children screamed at me as I made my way down to the back as if walking on hot coals. Inside the ambulance were wheelchairs fixed to the floor with little curtains around them and potties underneath. Everyone hurried to a chair and, not bothering with the curtains, we made a horrid chorus as the driver hurried, windows down, to Changi Hospital.

An Indian doctor was waiting wearing his turban and pleasant bedside manner. We wheeled into his waiting room.

'Gentlemen, I'll be needing a stool sample from each of you please.'

'There's some on my foot,' said Jack.

'There's some of yours on my foot too,' I said.

The poor doctor slid the potties out from under our chairs while we sat there, occasionally twitching, one by one going off to get cleaned up. The good doctor was bent over a microscope, then he sat back in his chair, spoke into a nurse's ear and turned to talk to us.

'We know what kind of parasite you have all ingested, so now we can administer the right medication.'

'Can I have a look?' I asked. I wanted to see what had done this to me. The boys curled upper lips and scowled at me as I wheeled over and craned my neck over the microscope. Swimming about in a frenzy were lots of tiny monsters looking like an underwater scene from *Jurassic Park*.

'Oh my God.' I felt like a victim in an alien movie.

'Yes . . . that is in your bowel,' he said, the calm way a doctor does.

'Give me the fuckin' shot.'

The upside of having had Amoebic Dysentery, apart from not dehydrating to death on an airliner, is that I haven't had to run to a toilet in years as the body builds up an immunity to parasites. The downside was, for months afterwards, I would be at a party or just standing around in the workshop, and when someone cracked a joke, everyone else burst out laughing, but I was sprinting to the toilet to check.

And I was scared to fart for a year.

There was a directional driller who I regularly ran into, a New Zealander by the name of Maurice. A good-looking man in his early forties, Maurice was very good at his job and very good at getting in all kinds of shit when he wasn't drilling. He was well respected by everyone, even when he got loaded and wild at company functions. Maurice needed a warning posted on his forehead, or explicit instructions for party holders to lock up their wives and daughters and organise a team of men with restraints to capture him in case he kicked off, as Maurice had no concept of fear, or of consequences.

Maurice told me about his farm in New Zealand and his dogs that he hunted wild pigs with. The best of the pig dogs were Chaos and Razor and from what I heard they, like their master, knew no fear. Maurice pulled a tattered photo from his wallet one day and showed them to me. I didn't know what to say. Chaos looked like a Staffy crossed with a rhino, and Razor was basically just a monster. Apparently Razor, in his youth, got hold of a massive pig in the mountains; the pig mauled him so badly he lost most of his jaw. Had he belonged to anyone else Razor probably wouldn't have made it out of the woods that day, but Maurice being Maurice carried the big dog home and drove him straight to the vet. The vet recommended Razor be put down, as without teeth he was unable to survive. Maurice, however, wasn't ready to give up and so had a dentist make, from scratch, a full set of surgical-steel teeth to be placed in the dog's recon-structed jaw. While Razor was waiting for his new teeth, Maurice had to puree all his food and hand-feed him, but the dog sank into a deep depression, kicked to the bottom of the pack in the farm's canine hierarchy; he was inconsolable. Until he got his new teeth. The result, much to Razor's delight, was terrifying. And he knew it. You could tell by the way the dog was grinning. He looked like he had a mouth full of chrome crab claws. Razor immediately ascended to the top of the food chain. These days, the wild pigs do a big double-take and run away.

10

THE DARKEST CONTINENT

2001

CHRISTMAS WAS ONLY A few weeks away, and my heart was broken. Pia ended it for a reason I had to accept, the oldest one in the book of failed oilfield relationships: I was away too much. I was completely intoxicated with Pia, the sun rose and set with her. When she left I was crushed. I moped around the house for weeks and weeks. The phone and doorbell would ring, I never answered them. Eventually the money started to run low, so I reached out looking for a gig but there was nothing going on.

Walking through town one Saturday night I ran into a friend, and he asked where Pia was, but for some reason I couldn't say the words. I just said she was doing her own thing. He offered to buy me a drink so we wandered into a seedy-looking first-floor bar in Kings Cross called

Barons. Hearing his news distracted me and I started to relax and enjoy myself. We had a few more drinks and talked about grabbing some dinner when someone dropped a coin into the jukebox and set off all my triggers at once. It was our song. I felt suddenly empty, as though I had just given too much blood.

He was recalling a skiing holiday in Aspen but I was hunched over my beer, unable to stop myself from crying.

He stopped talking. 'What's wrong Pauli?' He leaned in; the bar was getting crowded and noisy.

'I love Pia,' I said, speaking into my glass, too embarrassed to look him in the eye.

'I love beer too, mate, but I'm not going to cry about it.'

Christmas was only days away when I was approached with an offer to work in Nigeria for six months. I had heard so many bad reports about West Africa over the years that my immediate reaction was 'No thank you', but with a depressing Christmas looming and the lack of work in Asia, not to mention my state of mind, I phoned back.

I contacted everyone I knew who had worked in Nigeria and developed a morbid curiosity about the place. Once my gear was packed I was eager to go, and get away from Sydney before Christmas found me bah-humbugging my way through a bottle of scotch. The company emailed asking if I could stop over in Paris to discuss some 'logistical issues'. This was unusual, but I agreed as Mum and John were only a short train ride from Montparnasse Station and I could stay with them for a night.

I left Sydney very quietly. Ruby and Tony, good friends who kept my head up regardless of how much I pissed and moaned about my lack of a normal life, were there to wave me off. I was so depressed, but when I finally found my way to the seat on the aircraft and started thumbing through the in-flight magazine, my mood lifted for the first time in weeks. No matter what Africa had to offer, I knew it was a better option than skulking around the house over the festive season look-ing like someone just shat on the turkey.

The flight was your standard economy-class night-mare exercise in confined space anxiety management, with a couple of infants screaming so loud that dogs 35000 feet below us could hear them. I ate the mini-meal with the mini-plastic cutlery, watched the mini-TV trying not to get the headphone cord in my mini-mashed potatoes, while the overweight flatulent pensioner next to me made a concerted effort to have the whole cabin smelling like cabbage before the flight attendants had cleared away the trays. I'm sure everyone else in that section felt the same way. We were locked in, unable to escape because of the tray tables and the mini-food carts that effectively block your path to cleaner air and a mini-toilet where one could have a few mini-moments of relative solitude, or a claustrophobic panic attack, whatever gets you through today's budget air travel. Just think of the frequent mini-miles, not the deep vein thrombosis that will have you dropping dead on a beach on your first day.

Charles de Gaulle International Airport sucks. I had been through there on a few occasions and it was always a pain. Eventually I found myself standing in a heaving sea of tired passengers with methane intoxication, all jostling for a good spot near the luggage carousel. I could hear the two children from the flight still breaking the sound barrier and making people on the other side of Paris cringe and cover their ears. (It's not that I dislike children, I love them, and I totally understand it's the air pressure in their little inner ears causing them pain, and that they are too young to clear their sinus passages and equalise the pressure; it can be very painful, like flying with a bad head cold. I just don't have any experience with kids. Even with my sister's three boys, I am told I handled them like rabies-infected carrier monkeys. She made me change a nappy. I have seen men cut limbs off, all kinds of nasty accidents on the drill floor, but a nephew's shitty nappy made me gag and resemble someone trying to deal with an untidy parcel filled with a mixture of nuclear waste and velcro. But, as I've been told before, it's different when it's your own kid's poo, right?)

My taxi pulled up outside our Paris office and I was greeted by a friendly receptionist, who politely showed me into the manager's office. He was affable and smiled too much, and while he was offering me coffee I saw the 'poor bastard' look in his eye. The logistical issues turned out to be remarkably silly, almost comical if they were not so serious.

On arrival in Lagos, after going through customs and immigration, I was to look for my driver; he would be wearing green company coveralls and holding up a sign with my name on it. I would approach and speak this sentence and only this sentence, 'It's hot here, just like Australia.' The driver, upon hearing that, was to answer, 'Just as hot, but no kangaroos.' If he didn't say that, it meant that he had murdered the real driver, stolen the company car and was planning to drive me out of town, put two in the back of my head and make off with my stuff.

I was curious as to who thought up the dialogue, because it was bloody stupid. The French manager, puzzled at my sarcasm, said he constructed the sentences. I only hope one day I can see him arrive at Kingsford Smith, walk up to his driver and say, 'Bonjour, the potato is on the suitcase.' He didn't get it, and went on to emphasise that this was a vital protocol as two men had been killed exactly that way only a few months earlier.

So with my ridiculous cloak-and-dagger routine committed to memory, I walked off to spend an enjoyable evening with Mum and John.

We had a fabulous dinner and a special wine; we sat and talked until late about the entire goings-on in our lives. I ate too much *fois gras* and hoped my Nigerian driver wouldn't shoot me as I padded up the stairs to brush my teeth and sleep off John's three bottles of Château de Carrier Monkey. Mum cried at the train station the next day, and gave me her standard 'Don't forget your safety' line. But she and John knew the rigs well after thirty years in the business so it was pointless trying to gloss over where I was going.

The flight to Lagos was a carbon copy of the Paris flight, except I was the only white person. Babies screamed, the food was small and tasteless. Instead of a smelly pensioner I sat next to a middle-aged Nigerian woman who enjoyed telling me that Lagos Airport is the most dangerous airport in the world, and I should be very careful indeed. That's no problem, I thought, I'm just going to walk up to a complete stranger and come out with the dumbest line imaginable, and hope he doesn't blow my brains out for two hundred dollars in traveller's cheques, a Mars Bar and some dirty underwear.

I could smell Nigeria before I could see it. Then on our final approach I looked down and saw a mess that resembled a Manila shanty town after a typhoon, with extra shit and heavy on the random people with big guns.

Everyone disembarked, that is to say, everyone just got up and rushed the cabin crew, who flung the door open and let the passengers trample over each other in a bizarre scramble to get out. I thought, perhaps the flatulent pensioner from yesterday was on this flight? But when I rounded the last corner of the terminal I understood why: eight immigration booths and one guy on duty.

He spotted me a mile away at the back of the official queue-jumpers' queue, his eyes immediately bulging. You could almost hear the 'KA-CHING'. They told me in Paris to arrive looking poor, no jewellery etc., so I was looking lower than whale shit, and if you saw me at home you would cross the street to avoid me. But I could not disguise white skin; a bribe was definitely in order. I hoped they took traveller's cheques.

After dealing with him and his two mates in customs I was officially in Africa. Passing through large wooden doors I was suddenly confronted with a mass of black faces, all staring straight at me. Then just behind the front row of the crowd I saw the sign 'Mr Pauli'. Practising my line under my breath, I walked up to the fit-looking man in green coveralls who bore the sign and said in a loud confident voice, 'It's hot here, just like Australia.'

He gave me a blank look, then flashed a huge bent-toothed grin and said, 'I am de driva.'

Cocksucker, I thought, what do I do now? Leaning in, with lots of eye contact, I repeated, 'IT'S HOT HERE, JUST LIKE AUSTRALIA.'

'Oh yes sa, BUT NO HOT KANGAROOS.'

Close enough.

'I am Oscar de driva.'

'How do you do Oscar, now get me fuckin' out of here.'

'Very good Mr Pauli follow me, I have caa with air-condishanings.'

Oscar was only twenty-one and had already managed to father four kids. He had been working since he was eleven and looked it. His English was good, I especially liked his accent. He hoped we could talk a lot and perhaps I could help him learn to read and write. He was in the middle of asking questions about what it was like to live in a free Western democracy and sleep with white women when I became distracted by a large sign on the side of the road.

Fifty feet square in size, it read 'Welcome to Port Harcourt'. Most of the letters had long since fallen off, and the whole thing was riddled with bullet holes and covered in brown stains that suggested some strange explosion had occurred involving lots of hot coffee. Two huge black vultures perched on the top of the sign and a third was on the ground with his head buried inside what turned out to be the chest cavity of some unlucky Nigerian.

My jaw slowly dropped, as I tried to take it in. That summed up Port Harcourt perfectly.

Oscar flashed me another bent grin, palmed a pistol from under the seat and said, 'First time in Nigeria? No wahalla, you are always protected sa.'

Christ, for a second I thought he was going to point that gun at me, and I would be joining the vulture picnic, the second course in an ever-increasing pile of dead idiots. So I played it cool, only making him stop once so I could throw up.

Guns are as common a sight in Nigeria as mobile phones are in Sydney. In this respect the Nigerians put even the Americans to shame—but no, wait, guns don't kill people, people kill people, right? Oscar de driva always had his mobile phone and his gun on him. I thought Nokia should develop a camera/gun, or a phone/gun, or even a gun/phone/camera . . . there would be massive sales in West Africa.

Port Harcourt is a remarkably dangerous place. Once considered a jewel in the dark continent, war and corruption have destroyed it. And while 2.3 million barrels a day come from its belly, none of that is put into the most basic of human needs. All the oil service company personnel operate from bases surrounded by high walls with razor wire on top and armed guards on duty twenty-four hours. The sensible people only leave the confines of the base to go to a rig or to the accommodation that's within another secure base. If you must go out it's best done in a large group. All the transportation between these places is done with an armed driver and a guard who usually carries an automatic weapon and 130 rounds—hope that's enough.

We drove into town, past filthy ramshackle neighbourhoods and the kind of abject poverty that puts every creature comfort you have at home clearly in perspective. I was immediately thankful that I was born on the right continent. After a few miles I asked Oscar what 'No wahalla' means. He explained—it's basically 'No problem'—and for the next six months it was 'No wahalla this', 'No wahalla that'.

When we arrived at the base, Oscar beeped the horn and the two heavy steel doors swung open. The compound was 500 square metres of concrete with three large hangars on one side and three single-storey office buildings on the other. In the centre of the compound an ancient truck was parked and two security guards were pointing their rifles at a local man who was on the ground, cowering, his arms doing a frantic explanatory pantomime. He was talking so fast it came out in one long syllable.

Oscar ignored all this and pulled up in front of the largest of the administration buildings. I jumped out and ran over to the guards. 'Stop pointing those weapons at that man.'

They turned and smiled. 'Dis man was stealing diesel, boss.'

I looked at the accused man more closely; he was badly beaten, obviously suffering broken ribs, possibly haemorrhaging, going from his difficulty in breathing and the profuse amount of blood running out of his mouth. 'Who told you to do this?' I asked the guards.

The bigger one stepped forward. He had a cold look; he was enjoying his job and I wondered how many people he had killed. He pointed over my right shoulder but nothing was said. I turned and saw the angry base supervisor marching towards us. He looked straight through me and barked at the guards in a thick German accent, 'I told you to flog him well.'

The local man started crawling away slowly, under the truck.

The heavyset German had a round head and fair complexion, which in the African heat made him look like a bulldog that just swallowed a bee. He extended a sweaty hand. 'Hello, I'm Carl,'

'What the fuck's going on Carl?'

'Ya ya come, you don't know how we do things here yet.'

By the end of my first day in Nigeria I was disgusted to a point that made me feel ill.

After meeting all the expat crew stationed there, it was apparent that this base was a kind of Betty Ford clinic, as most of them were disgruntled middle-aged habitual alcoholics, who regularly entertained each other with a fist-fight. I celebrated Christmas and then the

New Year in the company of what's best described as lobotomised monkeys. They started out well, lots of handshakes and backslapping, but quickly degenerated into the kind of malevolent lunacy I thought was only a myth in today's oilfield. Right down to throwing full cans of beer skyward and unloading 12-gauge shotguns at them. This was followed by a rousing game of hurl-fireworks-at-each-other, the indoors version of course. I did get on well with a couple of them, but on the whole the crew was burnt out; for them the prospect of working on new projects with new equipment presented all the excitement of a blocked toilet to a plumber.

Luckily I spent more time offshore working on the rig than on the base. During one spell in town, the mechanic announced his new house was finished and he wanted to show me.

'I have been building it for almost one year,' he said proudly.

It was just around the corner from the base so I said we would go and have a look at lunchtime. When we got there I couldn't believe what I saw. There in amongst the rusty tin and mud bricks of a foul shanty town stood the mechanic's new house, constructed entirely of blue plastic milk crates bound together with baling wire. There were hundreds of them, with tin pop-riveted to the outside, forming the house. The whole thing stood about one foot off the ground on another series of milk crates. The floor was made of plywood from the packing crates used to ship our equipment. The toilet was a hole cut in the plywood floor in the back corner.

The mechanic's wife came out, followed by a procession of children. They lined up around the outside and with some help from the neighbours demonstrated how the house could easily be picked up and moved once the crap piled up underneath the toilet. He was already accumulating more milk crates to build an extension, possibly a nice porch or gazebo. Apparently the highest cost involved was the gun his wife was brandishing, because milk crates are highly sought after in the Nigerian building trade.

The crew's accommodation, a simple house which seemed palatial in comparison, came with its own set of problems. Getting home was one. For example, once when I was on my way back from a job, the chopper touched down just as the sun was starting to set, its rotor wash sending up a cloud of dust that smothered the small heliport. The driver and guard were waiting for me. That was a relief because on the last occasion I sat at the heliport for two hours, waiting, too afraid to take a taxi, because shooting me would be more lucrative than driving me to town.

As we drove through Port Harcourt I was told there had been a lot of rioting that week because of the elections. It was all very tribal, the Muslim 'Felani' in the north clashing with the 'Ebu' Christians in the south. When they get really pissed off with each other they wave limbs from banana trees over their heads, and if you see them throw the limbs then it's going to the next level. That's when everyone runs home and comes straight back with a gun, or a petrol bomb, or a machete.

The car was all over the place and it became apparent once he started talking that the driver was shitfaced. He took us on a brief but exciting detour into a lane of oncoming traffic, turning sharply down an alley and into the middle of a pre-riot banana branch-waving session. As soon as the crowd pinged my shiny bald white head they rushed the car.

Within minutes the driver was panicking; he stalled and flooded the engine and frantically tried to restart it. The crowd began to produce weapons and beat their fists on the windows. The Pajero rocked under the surge. I remember screaming at the guard to do something. He racked the cocking lever on the side of his AK, cranked down the window, stuck out the barrel, roughly pointing at the sky, and emptied the magazine.

Everyone shat their pants. The weapon kicked in the guard's hands, as empty brass casings spat across the inside of the car, hit the windscreen and glanced off directly into the crotch of the driver, who was still keying the ignition and pumping the throttle like the drummer of a speed-metal band.

Empty shell casings are extremely hot, and the driver was wearing shorts.

He suddenly shot up, banging his head into the roof,

but at the same time he somehow held the ignition on and bunny-hopped the car straight over a man who was trying to clamber onto the hood over the crash bar. We rocked over to the right as the car went over the man; I scrambled over my offshore bag just in time to see him emerge from under the car, one big Desert Dueler tread mark running across his flat body.

The guard had changed magazines and was hanging out of the window now pointing his rifle directly at the crowd, over his trigger freeze problem; he looked like he was enjoying the whole thing. We bolted flat out back to the base, and I made the journey laying flat on my face on the car floor.

Arriving back at the staff house, I relayed my story to the crew who just shrugged their shoulders as they had all been through similar things before; it's just part of working in Nigeria.

A few days later I found myself returning to the staff house alone, everyone was on jobs, so I sat down to enjoy a quiet night in front of the TV. We had a satellite dish on the roof and could get half a dozen decent channels, the house favourite being the cartoon channel. The boys would sit and watch cartoons through the night. The fact that they did that was disturbing enough, but what I found truly remarkable was that this channel carried commercials. What could you possibly sell to someone who voluntarily watches *Deputy Dawg* at three in the morning?

On this night I tuned it to HBO Movies: *Fight Club* was just starting, I hadn't seen it and was looking forward to it. Then halfway through the movie the wall socket behind me started making more noise than usual. Our electricity was supplied by a massive generator in the backyard that had come from some rig and was far too big for the house. The lights would regularly jump from 60 watts of glow to 100 and back down to 10, and the wall sockets would crackle and spark, giving the impression that someone was being electrocuted in the basement every half hour. But I was so used to it I just ignored it. Only moments later I could smell something other than Africa. Turning, I saw the wall socket, the air-conditioner, and a good deal of wallpaper on fire.

The fire extinguisher in the kitchen was empty; it had been used up during the New Year's Eve indoor fireworks display. I grabbed the small one from the car and put it out, but was worried the fire could reignite in the roof or wall space, so I checked all the rooms and found one of the guys asleep. He was most upset with me; not for waking him, but for putting out the fire.

'Let's re-light it bro and we'll get a new staff house.'

I talked him out of it.

Incidentally, the same guy would go jogging every morning; he was making a brave effort to do something about all the years of smoking and drinking. One morning I was standing at the main gate to the compound when he came jogging by. I was amazed that he was outside alone; he was taking an awful risk. But when I saw the reaction of the locals I understood why he had remained unharmed for so long. They would stop what they were doing and watch him jog by, then immediately look up the road at where he had come from: Who was chasing him? Why else would a man run down the road? He must be crazy, look at his red face, stupid white man.

Local kids would always loiter outside any place the expats went, especially bars. At ten years of age, they were already professional tiny hitmen. They would single out the weakest man, usually the drunkest, and handi-vac the contents of his pockets. In Nigeria, don't turn your back on anyone, ever, even if your pockets are empty . . . think of the worst crime, and they've done it, enjoyed it and improved on it. The security staff would kick the kids and chase them away. I found the best system was to throw a handful of change in the opposite direction of where you were going.

I was to discover later that adult thieves are dealt with more vigorously. The crew and I were in a mini-bus, driving across town on our way back from a job, when the driver asked if we'd like to see the public hanging at 2 p.m. It was 1.55 p.m. and we were a block away from the police station, how convenient. Outvoted,

I found myself sitting in the bus watching three shirtless, handcuffed men standing behind a roughly constructed gallows—I was kind of surprised it wasn't made of milk crates. Two police officers stood to the side, one wearing pyjama pants and a combat jacket, the other in combat pants and a pyjama top. The first two men died quickly, the rope snapping their heads back, breaking their necks instantly. The third man was at least 300 pounds and built like an ox; his neck did not break. We watched him thrashing about on the end of the rope and started yelling at the two cops to do something. They looked at each other and in a well-rehearsed manoeuvre laid down their guns, grabbed a leg each and pulled down, throttling the man to death.

But for all the public executions and floggings designed to create fear of authority, violent crime is an everyday part of life in Nigeria, the most corrupt country in the world. If there was something nice to say about the place, believe me I would say it.

The most outstanding event during my time there was to give me nightmares for months.

Late one night I woke to the sound of screaming, coming from the TV room. I thought I was alone in the house, but when I went to investigate I discovered one of our crew standing butt naked and holding a chair out in front of him rather like a lion tamer does. In front of him stood a Nigerian woman, obviously a prostitute, fully dressed and wearing a platinum-blonde wig. She was waving a knife and trying to rob him, having waited until he was naked before making her move. My appearance only made her more determined; now she wanted my money too.

'Pauli, help me man, this fuckin' bitch is mad.'

I turned on my heel and ran out the back, around the house to the front gate, where my highly trained, super-alert security guard was on duty, fast asleep against the wall. I woke him gently, tapping his shoulder and keeping my voice even and soft, not wanting to startle him in case he shot me. He looked up at my smiling face.

'Hi Daniel, would you like a nice cup of tea?' He knew I had done that for the other security guys as they were not allowed in the house.

'Oh yes please sa.'

'WELL TOUGH FUCKIN' SHIT . . . BIG WAHALLA IN THE HOUSE, YOU GET HER OUT NOW.'

Up he jumped. I followed him through the front door; blondie had our naked lion tamer cornered now. She was big, much bigger than me—imagine Mike Tyson in drag. As soon as she saw the guard, she dropped the knife and started babbling at him, but he didn't break his stride. He just stepped up and with his whole upper body swung the butt of his rifle into her jaw.

Both her feet left the ground, she slammed down hard on the tiled floor. Her wig flew off followed by a long arc of blood. I knew he had killed her, but my legs were frozen to the floor. We watched him drag her out, feet first, through the front door. A few moments later he drove off with her body in the back of the truck. The room boy casually walked in and started mopping up the blood.

I should have left Africa right then, but I had one more job to do.

Two weeks later everything was going just fine until all the roughnecks and rouseabouts, and everyone who wasn't white, walked off and came back five minutes later with weapons. Mutiny is the best way to put it.

There were sixteen expats on the rig at the time; the other eighty personnel didn't want to do any more work until they got more money. All the offshore workers are supplied by a government body called 'The Labour Mass' and they had decided to strike, with weapons, in an effort to force more money out of the oil companies. There were five rigs involved in this. The Labour Mass just picked a day and time, gave the men enough notice to smuggle weapons out to the rig and stash them, and on a predetermined day they made their move. They had control of the rig, ballast control, the radio room, well control, everything. They boomed the cranes over the heli-deck so choppers couldn't land or resupply the rig.

This went on for the next three weeks, one enraged man after another speaking on the radio, occasionally breaking for a barbecue and an impromptu chanting session on the heli-deck. The news networks were saying all kinds of outlandish crap, like herds of rig personnel were being locked into freight containers and dangled over the sea, sometimes even dunked into the sea so everyone inside was waist deep in water and in total darkness. No one was hurt, we just watched TV a lot. Eventually we ran out of food . . . that sped up negotiations.

Finally the US navy got involved. As soon as the words 'SEAL' and 'take the rig back by force' were mentioned, the mutineers dropped their weapons, saying, 'Okay we give up . . . can we keep our jobs?'

I got back to Port Harcourt and quit, jumping on the first flight home, never to return.

(11)

GOBBING

2002

I WAS BACK AT Louise's agency two weeks later; she had me working on a campaign for hair products for young women.

'But I'm a thirty-five-year-old bald man . . . I don't know anything about hair . . . it's been ten years since I last used shampoo!'

'That's why it's your project Pauli,' Louise said, and as luck would have it we aced it.

She helped me get organised and I enrolled in some courses in advertising at the University of Technology Sydney and I studied hard for the next five months. For the first time I had a sense of choice: I could do something with my life other than oil. I liked the study, I enjoyed the classes and exercises. I was happy to learn but not to compromise my lifestyle and inevitably I ran out of money. So I soon found myself in China.

The last part of the journey to China was the best. I got on the wrong ferry at Hong Kong International Airport: you're supposed to arrive at the airport and go straight to a ferry which will take you to any one of twenty ports in mainland China, where you get off the boat and finally get processed through immigration. Provided of course you get on the right boat to start with. What can I say, it's a rabbit warren in Hong Kong, so I ended up in a remote port.

I was already destined for an isolated province, but I arrived at a *really* isolated province. Even the immigration guy looked surprised. There were no phones, no taxis, or buildings, or other Westerners. And it was getting dark, although that's never bothered me. Darkness is your friend in dodgy remote Chinese ports where a tall bald white man in a Mambo T-shirt tends to stick out like chairman Mao at the MTV music awards.

So I had to fly via a local domestic carrier to the right town—not good.

China's answer to the global terrorist threat on airliners is simple. First they search you, then they search you again, then they give you the proper search, then in case they missed something during the proper search, they search you again, that's the search when they squeeze out your toothpaste, saw the heels off your shoes and X-ray your underpants.

Coupled with all the searching, you're bombarded with a kind of video-movie of what would happen if some terrorists decided to hijack a plane. Like a tribute to 'the golden years of terrorism', it features a man in black coveralls and a black ski mask in the middle of an aircraft fuselage who is brandishing an automatic weapon then gets shot a couple of hundred times by a small group of other men in black coveralls and black ski masks . . . which begs the question, did the right guy in black get shot here? Perhaps some sort of name tag is in order, or even a good pair of Kevlar comedy breasts.

So there I was on the plane, waiting for the all-singing, all-dancing black coverall-wearing ski mask appreciation society gala performance. Instead I ended up in an aisle seat next to an elderly Chinese gentleman, who must have been ninety and looked like he had built the whole wall himself. He had no teeth, which proved to be a bit of a problem when the in-flight mini-gruel was served. I did feel sorry for him because he had to deal with the joy of scoring my meal, combined with his meal, but he shook so much he dropped half his food in my lap. I would have been better off just up-ending my tray in my own lap and calling it a day.

After dinner I tried not to think about the boiled-gruel stains on my pants and my new friend decided to spit on the floor every five minutes, then he had to go to the toilet every ten minutes for the rest of the flight.

Either the old man hadn't had a meal in a few days or he had a prostrate the size of a nuclear submarine.

The town is called Shekou and feels palpably weird, like I wouldn't be altogether surprised to pull into a local bar and find my drink being poured by a Cyclops. And there's a real problem with counterfeit money, so every time I got change I scrutinised my bills like a diamond merchant.

I soon became quite used to people staring at me. Westerners are rarely seen in this part of China so it's expected that locals will have a good look. I should have been happy, all things considered: I wasn't getting shot at, not everyone in this country was carrying a machete, they don't hate white people and you can eat the food without that niggling feeling that there's a human hand in the stew.

The only thing I couldn't get used to in China was the gobbing. Everyone, and I mean everyone, hacks up a big ball of phlegm and spits it out on the street, every five minutes. Women, children, babies, monks, doddery old people who look like the next big gob could kill them—everyone has a good gob, all the time.

Perhaps the answer to China's economic problems lies not in oil and gas exploration, but in utilising its other natural resource: spit. It's a lot cheaper to find than hydrocarbons, all you have to do is set up millions of giant spittoons and find a way to convert the spit into some sort of industrial lubricant. They could spend the money on driving lessons for everyone, because when the locals aren't gobbing all over the place they are driving around like Stevie Wonder. (In China I came frighteningly close to getting flattened by anything from kids on rollerskates to rickshaws and semitrailers, but that's possibly because I was too busy trying not to step in all the gobs.)

My boss was a big gobber too . . . I think in his youth he gobbed for China. No problem, mid conversation . . . Whap! Right there on the floor, watch your step. Even the Chinese President has a sly gob in parliament; I saw him do it on local TV one night, God love him.

I try to enjoy local traditions and customs wherever I go in the world, so I decided to perfect my own gobbing technique and really impress the guys at the next drilling meeting.

Food is always another wonderful experience in a new country. China is crowded, all knees and elbows and gobbing, but apart from that the food is pretty good. Compared to Nigeria, it's the Ritz Carlton.

One morning's drilling meeting was especially fun. On this particular day, I met the tool pusher and driller who would be working with me offshore. They are with a Houston-based company and were on their first venture in this part of the world, real genuine redneck Gawd-damn American good-ol'-boys, with giant belt buckles and their bottom lips packed with two pounds of chewing tobacco. Coincidentally, these Texan boys enjoy a good gob too.

Watching their first meeting with the Chinese rough-necks was a treat; it could have been a MasterCard ad . . . 'Price of one round-trip ticket to China, $3000. Price of enough beer and cheeseburgers to keep you fat-n-stupid while in China, $5000. Finding out your entire Chinese drill crew also enjoy a good gob with absolutely no social graces . . . Priceless.'

Really good drilling people are allowed to be surly and indifferent about personal hygiene. But these guys were shocking.

We went out for breakfast after the morning meet-ing broke up. Redneck 1 says: 'I wanna eat what you boys eat.'

The Chinese guys exchanged blank looks and then all looked at me. I've seen this happen before, in other parts of the world, when expat oil workers try to experience the local culture with the best of intentions. I've ended up sitting down to eat in a backstreet dump with rats running along the rafters; I had to step over a goat to get to the toilet which was just a hole in the ground anyway. I shrugged at the Chinese guys and they shrugged back at me, but the rednecks insisted so off we went to a local food stall.

Redneck 2 takes one look at his traditional Chinese breakfast and says, 'Jesus Christ, I'm not eating this shit.'

Redneck 1 meanwhile has pulled off his genuine imitation crocodile Tony Lama cowboy boot and sock to investigate the galloping jungle foot rot that appeared to be eating his right foot off, displaying his exemplary table manners . . . and then the gobbing started.

To the Chinese, the Texans must have looked like something from a *National Geographic* special: Two huge hairy white men, with chrome hubcaps holding up their pants, and the blackest gobs they'd ever seen. Their street crèd went up instantly. If I had a flip-top head and managed to hack up an entire lung, I could not have topped those guys.

For the rest of that day we had a good look around the rig before it got towed out to location. While the rig had been in a dry dock getting painted and fitted out with a new drilling package, a stray dog came on board. I made friends with him and named him Colin; he liked me a lot and followed me everywhere, probably because Colin knew I was not likely to eat him. Besides I was nice to him and gave him lots of human food and for a dog in China that's a rare treat. I also made sure after seeing a couple of the boys eyeballing him that under no circumstances were they to eat Colin. I had the welder make Colin a little house that we put on the drill floor right next to the control station that the driller stands in, coincidentally called a 'dog house'. Colin was the only one allowed to shit in his.

Colin was a Chow, an interesting Chinese breed that looks a bit like a Pomeranian on steroids; powerful, but in a short compressed way. He was very dirty at first, covered in oil and all kinds of crap, but after a good wash with the steam cleaner and a proper blow dry with the high-pressure air line, he came out clean, with really big reddish-blond hair. He looked like an Asian David Lee Roth, except he really could lick his own balls.

Colin became the rig mascot and soon was embarking (no pun intended) on a cruise out to a quiet part of the South China Sea, hundreds of nautical miles from any major shipping lanes, where he was going to listen to his Van Halen CDs, drink out of the toilet, hump the furniture and not get eaten by the welder.

While I was hanging around Shekou I met Cameron, an aircraft engineer. Cameron was funny, in a slap-happy trigger-finger kind of way. A big heavyset character, ex-US Ranger, he was majorly into motorcycles and having sex with women he hardly knew. Every time I was in town I called Cameron and the games would begin.

Sitting in a downtown bar one night he told me about a mad driller who had been there only a month before. 'That Kiwi bastard belted the biggest guy in the room, kicked off the worst bar brawl in Shekou's history.'

'What started it?' I asked.

'Oh man, I think someone called the All Blacks a bunch of pussies and Maurice went bananas.'

'Did you say Maurice?'

'Yeah . . . a drilling guy, five ten, grey hair, maniac, from somewhere in the south . . . he's got these dogs man . . .'

I explained that Maurice and I were good mates and Cameron rocked back laughing. According to Cameron, the incident occurred in a biker bar and all Cameron's mates got into it with the giant Maurice punched out, who was the boss of a rival bikie chapter. It escalated from there to war. Cameron was thrown through a thin wooden wall into the bar next door. That started another fight, as people from both bars flooded through the hole, over the top of Cameron. The whole block erupted. Local Chinese police, too scared to come near the place, left them alone to murder each other. Maurice somehow got away. 'Next time you see that bastard, tell him the whole bar was destroyed.'

We had a good laugh, then Cameron suggested we go for a spin around town on his old Ural bike. The Ural is a Russian-made flat-twin thumper with a sidecar. Cameron took off with me in the sidecar, and we tooled around Shekou's dark streets. After a while we bumped into Dave, a mate of Cameron's. Dave was also drunk and decided to join us—he was so big he only just fitted into the sidecar—so I rode pillion. We had just taken off when we bumped into another of Cameron's drunk mates, John, who was in a wheelchair.

'There's no room ol' buddy,' Cameron said.

But John just grabbed onto the sidecar and we took off again.

Rounding a corner into the main drag through town, we passed a police car going the other way. He hit the siren and U-turned at some traffic lights. Cameron stopped but John was demanding we keep going.

'The chair can take it man.'

As soon as John let go of the bike, Cameron bolted, leaving an angry John curbside.

'We're in a police chase now . . . and my visa expired a week ago,' I yelled at him.

'It's okay Pauli, we do this all the time . . . The cops here can't drive for shit.'

He took the corners so fast that Dave's bulk was raised off the road. Finally we darted down a tiny alley that was barely wide enough for us to fit. The police car stopped, unable to continue. Cameron dropped me off outside my hotel. I sat on the front step and had a smoke. As I was butting it out, Cameron and Dave shot past on the Ural being chased by another police car.

RIG UP RIG DOWN

2002

I HAD FINISHED ALL my advertising courses while at home in Sydney, passing some well and others by the skin of my teeth. The day I received my final results there was an agency party for one of their larger clients that I attended. Happy in the knowledge that I didn't have to spend any more evenings at lectures or studying, I was in a great mood, but everyone at the function was even happier than me. Strangers were hugging me and talking the kind of bullshit that makes you want to move to deepest rural Australia and build a mudbrick house. After a few hours I excused myself, later realising that I was probably the only person at the party who was not on drugs.

In some respects the advertising world is just as bizarre as the oilfield, but there was no time to find out if

I was right about the party. The phone rang very early: there was a job in Japan, starting immediately. I jumped at the opportunity to go, as Japan is safe, clean and mostly bullshit-free. I had been to Tokyo briefly a few years earlier, and the people were as polite and as efficient as you would expect.

I had a two-day layover in Singapore then departed for Osaka, Japan, on a stormy Friday night. The job was twofold. First I had to go to a factory where some pipe was being fabricated, then to a rig up in the Japanese mountains. Osaka was much the same as Tokyo. They have the cleanest taxis in the world and the drivers wear smart uniforms with black hats and white gloves. They pull a lever to open and close the door for you.

There was a representative of the pipe company waiting for me at the airport. He chatted politely all the way to the hotel, organised everything, even what I wanted for breakfast the next day. I'm not used to that treatment; usually, once you walk out of the work-shop in Singapore, you're on your own. His name was Shouji, but I called him 'Turbo San' as everything the man did was at top speed. He would run to accomplish a task with the kind of urgency that left you wondering if he would get severely beaten or have a finger cut off if he was too slow. He had a complete itinerary that included, much to my amazement, cigarette breaks. He was so organised that I went to bed exhausted . . . and a little freaked out at having so much unexpected attention. I half expected him to spring from the wardrobe and tuck me in.

The next morning my breakfast arrived just how I like it. Turbo San was waiting in the lobby with another man who looked a bit like Eddie Munster with a bowl haircut. Turbo introduced us, Eddie was quiet, staying in the background, never joining the conversation. He wore a basic black suit, with a tie that looked like it had been made from my grandmother's curtains. I stopped giving him shit not too long after meeting him as I discovered that Eddie was a 'kendo' champion and could kill me with his big toe.

I asked if we could take the subway to the factory— a bizarre experience. Nothing in the station is in English and every level is a carbon copy of the last. We went down crystal-clean corridors past tiny neon caves where Japanese hawkers sold tiny perfect meals to tiny perfect clones. Everyone is basically dressed the same, though occasionally, however, you see the odd punk, straight from a Sid Vicious production line, complete with 'Never mind the bollocks' T-shirt and safety pin through the nose.

As we waited for our train, Eddie walked over to a huge bank of vending machines and got himself a paper. I noticed one of the machines was getting more attention than the others. So I had a look. The other vending machines sold every kind of drink, snack, cigarette or consumable imaginable, but this one sold something only the Japanese would want. The machine itself was cool; very retro with big chrome knobs and dials. It had a large glass window which displayed Polaroid photographs of schoolgirls in pig tails, with their name and age underneath.

When fed money the machine spat out a 'soiled' pair of knickers with bunny rabbits printed on them in a zip-lock clear bag, with a dirty letter to you from said child, complete with a lipstick kiss at the bottom. Turbo looked embarrassed, and laughed out loud as the Japanese do when embarrassed.

'Pretty girls,' he said, beaming.

We left the porno Wurlitzer and got on the train, which was so clean it was like riding in a long thin hospital. It stopped, lining up perfectly with markings on the platform floor. Everyone got off without any of the shoving and pickpocketing I was used to.

After our 'immensely serious' working day had ended, at five o'clock exactly, the senior man in the room and director of the pipe company stood up and announced to a packed boardroom that we would meet in an hour for drinks and dinner. Everyone then bowed to each other in a polite headbanging session for half an hour.

For the Japanese, work and play is like flipping a big switch. We had only been in the bar for ten minutes when the director jumped up onto a coffee table and with microphone in hand belted out a Sinatra number, much to the delight of everyone.

I was asked what I would like to drink; given this crowd's expense account, I requested my favourite single malt and was assigned a small woman who had a whole bottle of thirty-year-old Macallan. She didn't speak, just poured and went for ice. Ten minutes later I was on the coffee table, with my seat cushion up my shirt, doing my best bloated near-death Elvis version of 'My Way', complete with the belching and sweating. They clapped like lottery winners; Turbo hugged me with arms as broad as the *Enola Gay*.

We all got ratted and then went for dinner. I had Kobe beef, easily the best steak I've ever tasted. When I heard how much it cost I asked if they got a free TV set with that. The cattle which provide the heavenly steak are called Wagyu and come from Kobe just down the road. They are raised on a diet of grain and beer, and they get massaged daily. It's hard to imagine such a profitable commodity based on a grain-munching, beer-swilling animal, unless you've had experience with large multi-national oil companies.

The next morning I took the Bullet train to Tokyo and spent the day drooling in as many custom-motorcycle shops as I could visit, then continued up to the town of Nagaoka, 260 kilometres north, which was near the rig. Supersonic concrete gave way to a blur of beech trees through my squeaky-clean train window, as we left the high-rise metropolis which sprawled along the south coast.

Nagaoka was a world away from the industrial epicentre I left behind. It resides in quiet tradition that's guarded closely. Its rolling green hills open up to snow-capped mountains. The town is split in two by the Shinano River that flows from the nearby Higashiyama mountain range. A vast patchwork of rice terraces was peppered with stooped old farmers, who would always smile happily, unfazed by an inquisitive *gaijin*, or 'outside person', wandering about in the middle of nowhere.

Nagaoka is the birthplace of Admiral Isoroku Yamamoto, the Harvard-educated naval strategist who planned the Pearl Harbor attack. Apparently he never wanted to go to war with the US. By intercepting and decoding a secret Japanese transmission containing Yamamoto's itinerary, the Americans were able to shoot down his transport aircraft over the South Pacific in 1943, two years after Pearl Harbor. It was the first time the US succeeded in eliminating a major enemy leader by direct attack.

The rig was a thirty-minute drive from Nagaoka, at the foot of a majestic valley. It looked like a wart on the face of a centrefold, but it was the most organised land rig I've ever seen. Everything was done just so, and the crew started their shifts in company colour-coordinated starched tracksuits doing sit-ups and exercising in perfect unison. They have a work ethic that puts the rest of us to shame. The job went perfectly, they even cleaned our equipment for us when we rigged down.

I had two more days of exploring in Nagaoka before my train departed for Tokyo and home. On my last morning there was an earthquake, another one; I'm just lucky, I guess. We were standing outside the train station eating rice cakes when everything started shaking. People exploded out of the station like shrapnel from a grenade. Like Californians, the Japanese are used to earth tremors except that they can run in terror and still make it look civilised, you know, without the looting and trampling of children. Nagaoka's 200 000-strong population lost thirty-one that day with 3400 injured. It was a 6.8 quake, the deadliest since the 1995 quake in Kobe, when all those expensive overweight drunk Wagyu cattle must have made a mess. I went home to quake-free Sydney for a week of cheap steak and looting.

That week in Sydney passed so fast that when I got back to Singapore I thought I'd dreamt the trip home. Sometimes I move around so much I wake up in a hotel somewhere and it takes a few moments to remember where the hell I am. I've caught myself about to pee in the closet more than once in a dark hotel room.

13

LEGLESS IN RUSSIA

2003

SINGAPORE IS WHERE MOST oil service companies base their Asian operations. There is a massive industrial complex on the island's south-east coast called 'Loyang Offshore Supply Base'. It's from there that everything starts. I had been freelancing on a day-rate basis for the last four years, so wherever I went in the world, it began and ended in Loyang.

Singapore itself is basically one giant island shopping mall; all the crews who constantly pass through the area tend to congregate in the same places. Namely Orchard Towers, a four-storey building downtown that houses seedy bars from top to basement, affectionately called 'Four Floors of Whores', with some of the most un-imaginative names in public house history such as 'The

Bar', 'The Pub', 'The Beer House', and full of drunk rig hands and Filipino prostitutes looking for a meal ticket.

The word from Loyang was that there was enough work kicking off in Russia to keep me going for two years. The first job was on a mono-hull rig offshore for Shell, and later BP were planning on drilling a wildcat well, or exploration well, in the same region probably with a semi-submersible rig. All this drilling was to take place off the coast of Sakhalin, an island peninsula running some 300 kilometres up Russia's north-eastern seaboard and ending in the Sea of Okhotsk. Sakhalin is predominantly a flat tundra with only a few hills dotting an otherwise harsh landscape, and winter temperatures that plummeted to minus 60.

We would be flying via Seoul, Korea, to Yuzhno, the biggest town on Sakhalin and located on the southern-most point of the peninsula. From there we would take a train all the way to the top and a town called Nogliki where a camp was in the final stages of construction. Choppers would ferry men to the rig from there, especially in winter when the sea froze fifteen feet thick around the rig.

Hearing about all this in an airconditioned office in Singapore was surreal. As usual, I had no idea what I was in for. I went through the normal preparation for a new location but I couldn't imagine what minus 60 would feel like. The rest of my crew arrived in Singapore over the next few days; only one had been to Russia before and he said in winter you could go outside with a hot cup of tea, throw the tea up in the air and it would land frozen solid on the ground.

It's a constant battle trying to get your gear and logistics sorted out, always the cheapest option. Unfortunately most of this is organised through a management system run by personnel who either have never worked in the field or haven't been offshore in twenty years. But they regularly stand around crowded oilfield bars telling stories about the time they put out the fire, drilled the well with one hand tied behind their back, got the chopper just in time, and fucked Playmate of the Month in the back seat of the car on the way home.

A few days later we were on a plane and picturing attractive buxom Russian girls mincing about in the snow in furry bikinis. Collective daydreaming shut down when we boarded the Soviet-made airliner in Korea. Next stop Mother Russia.

The flight attendant looked like the offspring of Boris Yeltsin and Eva Braun, and the seats had not been reupholstered since the mid-1970s. The whole aircraft looked like Ken Done had thrown up all over it. Instead of having individual seat pockets containing the emergency card, sick bag and in-flight magazine, there was just one great big photocopy of the emergency card nailed to the wall next to the toilet. Someone had written something in ballpoint pen under the picture of a man demonstrating the crash position: 'In the event of an emergency landing do not attempt to suck your own penis.'

After a while the captain crackled to life over the PA system. He sounded British and was definitely in a bad mood. 'Yeah this is the captain. I'm up here on the flight deck with first officer . . .' SNAP . . . SNAP . . . We could hear him clicking his fingers at the co-pilot, who fired off his surname loud enough for everyone to hear. 'Right . . . We've levelled off at our cruising altitude of . . . SNAP . . . thirty-odd thousand feet and ah . . . if you look out the window on the right side of the aircraft, you'll see a big wing.'

Nice one, I thought.

Yuzhno is rough. There were thirty-seven murders, not to mention eight bear attacks, in the month when we arrived. It has the highest crime rate in the entire Russian federation. It helps to be an alcoholic just to live there. Our contact was nice, but he looked like he was on the local wife-beating team. Eighty years of communism don't just disappear and in this part of Russia you could be forgiven for assuming Perestroika had just happened.

Getting processed into Russia is quite an ordeal; there's a lot of queuing with people who look like they're auditioning for *Schindler's List*, as well as some very novel baggage-handling techniques. We waited two hours for our bags to travel fifty metres on a trolley being pulled by a tractor that looked like it had just finished ploughing a field. Then the bags, smeared in mud and cow shit, were hurled through a hole in the wall. It was like going back forty years, with bad fashion and lots of vodka.

We piled into a big 4WD and drove to our agent's office where he gave us border passes and train tickets. The train was like something out of an old war movie, with wooden carriages, lots of smoke and billowing steam. The journey up to Nogliki was going to take sixteen hours; we had fairly nice compartments with a heater and comfy beds. I shared my compartment with an American driller named Bobby; it was his first time in Russia too, and so far he was loving it.

Just as we pulled out of the station a guard came into our compartment. He was wearing a big furry coat, a huge hat with ten pounds of brass macaroni on the brim and a shiny AK-47 rifle. It doesn't matter where you go in the world, the two things you're guaranteed to see are Coca Cola and the AK-47. The guard explained that the whole train belonged to the company we were working for, and we were not to leave this carriage under any circumstances. We even had to sign a form to this effect, and on departing he turned and said, 'No vodka da'. We both grinned and nodded.

Two hours later Bobby was getting bored. The window was steamed over with condensation, and I had my nose buried in a book, so he got up and announced he was off to find some Russian dudes to talk to.

'You can't go into any other carriages Bobby.'

'Fuck all that. C'mon.'

We carefully opened the sliding door and stepped into the long narrow corridor. The motion of the train rocked us to and fro and every now and again we had to hold on tight to stay on our feet. Bobby opened the outer door of the carriage; the wind slammed it back against the wall. Ice covered the hand rails and the wind bit into my face. We were standing on a small grated platform, the frozen track whistling past below us.

'Jesus, Bobby, fuck this,' I said. We had to jump over the train's carriage coupling in order to get to the other carriage.

'Come on man, it's just like a Western' and with that he was over, on the opposite grating, thumbing at the door lock.

I jumped over and huddled next to the door; the cold was starting to freeze me. Bobby got the door open and we stepped inside the carriage. I slammed the door shut and turned around to see about twenty Russians, all motionless, mid-conversation, staring at us. Everyone was smoking; the air was thick with the stench of foul Russian cigarettes, BO and vodka.

Bobby raised a hand and beamed. 'Hey fellas.'

'Just like a Western mate,' I said and elbowed him in the ribs. All I could hear was the track's rhythm and the awkward silence. My brain was frantically searching for the Russian 'Hello' but I was blank.

Someone yelled out from the back of the carriage, but the smoke was so thick you could not see that far. We could hear a weird squeaking sound near the floor coming closer. Then out of the blue smoke a man on a small wooden trolley emerged; he had lost his legs well above the knee, and was propelling himself along with his gloved hands. The trolley looked homemade, using furniture castors. There was a huge bottle of vodka wedged in the man's crotch between his stumps, and a small silver cup dangled from a chain around his neck.

He pulled up at Bobby's feet, pointed at the two of us, yelled some more abuse and punched hard up into Bobby's crotch.

There was that weird two-second pause which happens when something hits your balls. Bobby's hands shot down grabbing at his crotch, he spun around, his face distended as if about to sneeze and then the pain hit him. It was like watching a big tree fall over. The legless Russian bobbed up and down on his trolley with excitement. Two men next to me burst to life, so I backed up, fumbling to open the door, but they grabbed the legless man, barked at two guys near the window, who opened it, and threw him out of the train . . . trolley, vodka, the lot.

'Fuckin' hell, Bobby, get up, get up.' I was back at the door, my right leg vibrating at the knee with adrenalin:

it's a condition called St Vitus's Dance, and I get it when there's going to be a fight, or if I have to speak publicly.

To my surprise the two Russians helped Bobby to get up and walked over to me, showing the palms of their hands as if demonstrating there was no threat. Then the train stopped suddenly and the guard from earlier came storming in. It appeared everyone on the train was blind drunk, except us. The guard yelled at everyone in high-speed Russian, but after a few minutes he put on his big furry coat and went outside with two others to retrieve the legless man. The Russians told us that this happened every time there's a crew change: Trolley Man makes a prick of himself . . . and they throw him out the window.

The train's path through the flat landscape is on raised tracks, and it never goes faster than about twenty kilometres per hour. In winter the snow drifts on either side are deep enough to cushion Trolley Man's plunge. He just sits in the snow drift freezing his stumps off while the boys stop the train and carry him back inside. He spends the next few days building a new trolley; he even steals the little wheels off the furniture from rigs in preparation for another window exit.

I thought it a bit much, treating Trolley Man like a crash-test dummy. But the Russians explained, 'He was in war in Chechnya, many here from war . . . he likes go out window.'

Apparently, almost the whole crew had come from

the Ministry of War. Trolley Man had stepped on a mine, and the government had rewarded him with a new job as a radio operator on the rig.

Bobby and I sat in the Russian crew's carriage and drank vodka. They had to make the journey on wooden benches and we felt quite ashamed to have our own private cabin with heating and beds. But that's the way it is, the Russians didn't care, most of them lived in tiny one-bedroom flats without hot water or a phone. For them a rig is virtual luxury, with hot showers, good food, they don't have to worry about getting their brains blown out for Russia and it's all free. They are hard men, good workers, I even took a shine to Trolley Man—he may have been an abusive violent alcoholic, but any double amputee who enjoys taking a dive out of the window of a moving train is okay in my book.

By the time the train pulled into Nogliki Station, Bobby and I had sobered up but had massive hangovers. An army truck was our transport to the camp, which was basically just a series of portacabins all joined up, with a high fence surrounding them and twenty-four-hour guards patrolling the perimeter. I think the security was more to keep us in than anything else.

My cabin was comfortable, I shared it with Peter: he had been with the company for a long time, and he has pissed more blood, drunk more beer and fucked more bimbos than anyone, and, oh yes, he's never been sick at sea. (A week later we were on a six-hour crewboat ride to the rig and Peter spent the whole trip on all fours vomiting.) Peter's a real character and always has a story that leaves you feeling like you just got all the enamel peeled off your teeth. But he's good at the job and I never had a problem with him.

We were on board the rig for a month, the job went well and my crew change came around. After a month offshore you're really looking forward to going home. I had done back-to-back jobs, so all I could think about was getting home to Sydney.

I spent a day back at the camp, waiting for the train. The sixteen-hour trip was a quiet one as there were no Russians this time. The camp had given me a packed lunch and a bottle of water, so after a few hours I decided to eat. The first bite was the last—as pain shot up my jaw. I spat out the contents of my mouth on the floor and jumped up to look at my teeth in the mirror. One of my back teeth had an abscess, I thought, it really hurt. By the time the train arrived in Yuzhno thirteen hours later, I looked like I'd jammed a cricket ball in my mouth and the pain was excruciating.

My agent was waiting there to drive me to the hotel because my flight to Korea didn't depart until the following afternoon. Running up to him, I saw his expression change. He asked if I'd been fighting.

'Get me to a dentist . . . dentist . . . dentista . . . tooth doctor, understand?' But he looked confused. 'It's my fuckin' teeth man.'

He nodded finally, pointing at his front teeth, 'Da, da, dentist.' He smiled, put a hand on my shoulder and announced, 'Today is Russian public holiday.'

'Oh fuck off . . . I need drugs then . . . pharmacy . . . pharmacy . . . chemist . . . drugs.' I couldn't see straight I was in so much pain.

'Okay we go.'

He took me to the local hospital then he did some fast talking to the girl at the counter, who looked over her shoulder at me. I was sitting on the floor, twitching.

Then the dentist appeared in front of me. I couldn't believe it. Russia's supposed to have good dentists. His white coat had blood splattered all the way up his right lapel and there were Cyrillic tattoos on his knuckles. This is bullshit, I thought, as he walked me into a barren room. I took one look at the selection of tools decorating a small table and walked out.

My agent then took me to a chemist, and I immediately started munching on painkillers until I passed out. I woke up in the hotel the next day, and just for a second I thought I was okay, but it was a fleeting second—oh this is bullshit! The phone rang; the agent had managed to book me on the early-morning flight to Korea.

I had a brief, painful layover in Seoul where I threw down more painkillers. I checked the bottle, the Russian chemist had written in English on the label 'Only four a day'—fuck, I'd had double that already, but at least the pain was starting to ease off. I couldn't stand the thought of an eight-hour flight to Singapore, but it was the fastest way to get to a dentist that would do an extraction without using a chisel.

The pills began to really kick in as I was boarding the plane. I was fast approaching a vegetative state, but I stayed just conscious enough to make it to my seat. The elderly Russian man sitting next to me thought I was handicapped and I discovered that I'd been drooling all over myself. He offered to help me get to the toilet; unable to talk I shot him a filthy look, but it must have seemed more like a cry for help because he called the flight attendant and told her that I was a handicapped gentleman who needed some assistance. I was just trying to focus on my seatbelt, when the attendant came.

She was your typical Singapore Airlines girl, very pretty, very small, so off she went to get help. Two more small pretty girls showed up and they talked about how come they didn't know there was a handicapped man on the flight who had special needs. All I could do was grunt; I was so trashed, my head was so swollen, everything just came out in one big syllable.

'I'm-nut-handicapped-is-my-tooof.'

She smiled. 'Yes, okay, Mr Carter, this way.'

I could barely walk, and I practically fell into the toilet cubicle. I tried to smile, she smiled back and slammed the door shut. I looked at myself in the mirror. A long string of drool was making its way down my shirt, my head looked disfigured . . . freaky. 'Wow you look really bad,' I told myself. 'I think you have an abscessed tooth, you can die from an abscessed tooth.' And that was really funny, so I sat there and had a good laugh, then attempted to urinate.

Once I'd lined myself up with the bowl, I let fly, but the aircraft hit some turbulence and I pretty much just peed all over the place. That was really funny too, and I laughed out loud, because there wasn't any turbulence at all, I was just fucked. I have never experimented with drugs, other than grass and that was ten years ago. Not because I didn't want to but because we are randomly drug-tested at work. This was different from any drunken state I had ever been in, I was lucid but totally spazzed out when it came to hand–eye coordination or talking. At least there was no pain; I'll take hallucinations and peeing on myself over that pain any time.

As happy as I was to amuse myself in the toilet all day, the flight attendant came back.

'Mr Carter, everything alright sir?'

I tried to open the door and talk, but it just wasn't going to happen. This is when I discovered flight attendants can open the toilet door from the outside. She looked at me with a mixture of pity and contempt.

'I-doynt-evn-like-taaaking-aspeerin,' I said.

She explained that I had been in there for more than an hour, and I should go with her now to a special seat. I wanted to tell her I was not handicapped, and that

I reserve the right to pee on myself for special occasions only. But she was talking to me like I was a child.

'Thish-is-sush-bullshit,' I said.

She sat me down at the back, I had a whole row to myself. Then she appeared next to me with my pain-killers in her hand; I must have dropped them. She asked if I needed to take one.

'Nooooo-ooo,' I said and so she gave me my mini-meal. Embarrassed enough, I shifted over to the window seat and tried to negotiate the mini-food onto the mini-fork and eventually into my mouth. I had not eaten since I left the camp, and that was two days ago. It's okay, I thought, I'll just chew on the left side, but getting the fork lined up with my mouth was more difficult than I had expected. I was dangerously close to taking out my eye. Finally, the angle looked right, and I jabbed the fork directly into my abscessed tooth.

'AAAAAAAAH!'

So then I used my fingers instead . . .

The flight attendant appeared next to me again. She tried not to frown at the grown man who just pissed all over the plane and was now playing with his food. She asked again if I needed a pill.

Our descent into Singapore airspace was like having my toenails pulled out with a pair of pliers, and I was in tears by the time we got to the gate. The pills had worn off; in lieu of being handicapped I had regained the power of speech and explained what had happened to the cabin crew, who kindly let me get off first and arranged for my bags to go first as well.

Every step from the airport to the dentist hurt. Our workshop manager, Joe, is an ex-offshore man and thoroughly reliable and he had arranged everything for me. A car was waiting and I went straight to an excellent dentist.

The dentist looked excited to see me. He took X-rays, rubbed his chin a lot while he studied them, then looked at the pills I had taken. 'How many did you take?'

'Eight, I think.'

'Eight! Mr Carter, that was extremely dangerous.'

The dentist then explained that I had two options: I could book in for surgery, which he recommended, or he could do the extraction now, however, there was a chance I would lose the sense of touch in my lower lip permanently.

'Just fuckin' take it out now.'

Twenty minutes later the tooth was out, I had a mouth full of stitches, more pills, and a quiet hotel room waiting for me where I could sit and drool in peace. The dentist said that when I woke up the next day I should hopefully have regained my sense of touch in my lower lip, and thankfully, I did.

I spent the next five months in Sydney, with one trip to France to visit Mum and John for two weeks. While I was there I rented a car to go exploring. Mum and John lived in the Dordogne region, which is predominantly rural and full of picture-postcard villages that made me want to wear collarless shirts and baggy pants with braces.

One day I was driving through the rolling hills going nowhere in particular when I saw a well-dressed man strolling through a grassy field with a big bird on his arm. I stopped the car and hopped over the fence; he saw me coming and smiled, motioning for me to come over. We had a polite greeting, almost formal. He had a beautiful hawk perched on his gloved hand, a basket over his shoulder and a white ferret in the pocket of his tweed jacket. He was hunting rabbits.

'Please have a seat.' He pointed at a blanket near the fence.

I watched him pull out the ferret, whose name was Claude, from his pocket. As soon as Claude hit the deck he was off down a rabbit burrow. A few moments passed and then panicked bunnies took off across the field, their bums streaking a white fluffy blur in three different directions.

The hood was pulled from the hawk's head; it lined up the nearest target and was airborne in seconds. The bunny didn't stand a chance. I had no idea that rabbits scream when they die; it sounded like a child. The hawk held the rabbit in its talons until the Frenchman strolled up, then it returned to its perch atop the gloved hand. Claude came lolloping up and sat next to the man's leg, patiently waiting to be picked up and returned to the tweed pocket.

I can't remember the hawk's name, but going from its ability it should have been 'Death from Above'. What is it about birds of prey? They know they look cool. Anyway, at least it was more civilised than blowing the bunny's head off with a shotgun.

14

THE GHOST OF A FLEA

2004

I HAVE A FRIEND in Sydney who has a pet ferret named Freddy. We go out riding our motorcycles together, and Freddy comes along too. He's an experienced passenger and just curls up inside Andy's backpack and falls asleep. Andy could strap Freddy to the handlebars if he wanted, and he wouldn't wake up. When I first discovered Andy was riding about with a ferret in his bag, or sometimes stuffed in his jacket, I was worried the ferret might jump out or bite Andy. But over time I learned that ferrets have a defined lifestyle. They sleep for an hour, then go nuts for half an hour, back to sleep for an hour and so on all day. So if you go for a ride with a ferret in your pocket, make sure it's after they've had the half-hour of going nuts.

Naturally curious and interested in anything they can climb into, a ferret in a new room is very entertaining. We would pull up outside a pub, go in, order a few beers and rack up the pool table. Andy would pull a totally limp Freddy out of his jacket. Freddy would wake up just in time for the cue ball's crack into the pack. Andy calls it Freddy Pool: the balls spin across the table, and soon Freddy is into it. He loves pool, and quicker than you can say 'No ferrets on the pool table' he is off down the nearest pocket.

After a couple of minutes, Freddy would spring from a random pocket, scarper across the table and down an opposite pocket. This went on throughout the game, but you had to be careful not to play for more than half an hour, as the little shit would just fall asleep somewhere inside the pool table and then you had to wait for an hour until he woke up. This happened on one occasion.

We were in Wollongong playing 'Freddy Pool' when he fell asleep in the table. No problem, it was early afternoon, in the middle of the week . . . then four bikers came in and started playing doubles for money.

Andy and I sat there, waiting, middle of the fifth game, when suddenly Freddy took off over the tabletop.

'Fuck . . . hey, did you see that fuckin' thing?' said one of the bikers.

'See what?' said his mate.

'There's a fuckin' rat in the table.'

'Don't be fuckin' stupid, Macca.'

'I'm fuckin' telling ya, there's a fuckin' huge fuckin' rat, inside that fuckin' table.'

'You cunt . . . anything to cop out of twenty fuckin' bucks a game.'

'You sayin' I'm a fuckin' liar, Davo?'

'Next it'll be . . . Sorry boys, can't finish the game cause a fuckin' emu flew in and stole the fuckin' cue ball.'

The two men began to shape up to one another when Freddy stuck his head out of the corner pocket, wondering what the hold-up was.

'There's the fucker, get it Davo!'

Davo was amazed, he just stood there, his mouth slightly ajar, while the other three bikers started laughing. Macca was intent on bloody murder; he hovered over the corner pocket brandishing the cue over his head.

Freddy popped out of the opposite corner pocket, ran into the middle of the table, did a nice little figure of eight and disappeared back down the same pocket. The cue came down hard, completely missing Freddy and splintering on the table's edge, sending fractured wood in all directions. Now all three of Macca's mates were folded up laughing. The bar manager came over with a security guard to calm him down. Macca threw twenty dollars on the table for the broken cue and stormed off, followed by his mates who were still laughing.

'There really was a rat in the table, mate,' they said to us as they picked up their helmets and gloves. A few moments later we heard the big 'V' twins fire up and roar away.

'Can we go now, I don't want to get my head kicked in over your ferret,' I said.

'Oh shit,' Andy said, looking at his watch. 'He's gone to sleep again.'

'Jesus, Andy, is the fuckin' thing narcoleptic? Wake him up!'

Andy got the keys to the table from the barman who told us that it was the best laugh he'd had in ages. We opened the table, retrieved the sleeping Freddy and made for home.

During that trip home I decided to change laundromats. The previous couple I had been going to changed the colour of one too many shirts, so I wandered into a slick-looking new one down the street. All I could see behind the counter were legs and the best-looking bum in history. Her back was turned, giving me a chance to take in her figure. She turned, and I was caught. Dazzling smile . . . stunning. I felt like a twat, she was so sexy, her cheeky enthusiasm toyed with my embarrassment, boggling my loins to a point that left me instantly unable to talk.

I went back the following Wednesday, at the same time, and there she was.

I found myself feeling butterflies every time I went to get my laundry done. I tried to look cool; I'd park my bike in front of the door, toe out the kick-stand and walk in grinning like a lottery winner. 'Hi, how are you today Paul?' Wow, she remembered my name. But I mostly fumbled with my backpack and came out with hope-lessly inane conversation. But she always flashed me that smile and said, 'Have a good day.'

That went on for months. Her name was Clare. I would look forward to laundry day and another chance to fuck up my thirty-second window to ask her out. Then one Wednesday she said she would rather be with her family as it was her birthday.

'Oh, many happy returns.'

I would have been better off saying 'Sucked in!' Many happy returns! Jesus, Pauli, that's hip and youthful. So I walked down the road and bought a big bunch of flowers. She was happy to get them, and as no one was in the shop we chatted for a while.

'What is it that you do for a living?' she asked. 'I'm curious because you can tell a lot about a person from doing their laundry.'

'Really?' I said.

'Oh yes, for example you're single, you only wear something once, your clothing labels are from other countries, so you travel a lot, right?'

'I'm impressed.' I was, of course.

'You disappear regularly, and your clothes have

numbers written on the tags when you get back, so it's got to be something like the merchant navy or mining. Am I close?'

I explained that the numbers are room numbers, that I work on the rigs and the laundry guy always writes your room number on your gear. 'You're right about everything else too.'

She smiled. 'So where was your last job, somewhere interesting?'

'Russia, Japan, and before that Africa.'

She thought I was lucky, travelling so much; she had just returned from a trip to South Africa. 'Whereabouts in Africa were you?'

'Nigeria.'

'What's it like?'

'Oh, it was definitely interesting.' I had to ask her out, but she'd seen my undies. What if she was an environmentalist and decided that I was nothing but a meat-eating eco-vandal who raped the Earth for a living?

'Would you like to have a coffee with me?'

We had coffee the next day. I was happy; it had been a few years since I had felt that way. Clare was wonderfully easy to talk to. It felt different, she was different, strength of character hovered under her features. All my previous relationships had ended because I was away too much. Perhaps this would be different. These emotions are hard to fathom, especially after years of listening to hundreds of men sit in the locker room offshore going on and on about their divorces. Here are a few quick snippets of oilfield marriage advice:

'Don't do it.'

'Cheating bitches.'

'She took the kids and the fuckin' dog.'

'Hide your money, man.'

'Fucked the whole team while I was offshore.'

'Got home to an empty house.'

'I'm gonna have her knee-capped next month.'

'So I fucked her sister.'

Etc., etc.

It's simply fear, I think. I always had this notion that I could just roam the planet and run pipe, get into adventures, continue fucking about like I had since I was in my early twenties. Meeting Clare made me think about the next ten years—shit. Basically, I'd have to pull my head from my arse and take stock. What had I done over the last fifteen years? All my friends are genuine grown-ups; they have mortgages, kids, and jobs that don't involve making conversation with 'Billy-Ray' during a typhoon about fuckin' turkey season back home Gawd-damn.

There was one person I could talk to about this: Ruby. The constant in my life, my oldest friend, we go way back, nothing was real until I told Ruby. She always threw new light on everything. I had been threatening myself for the last few years with the idea that I could do something else with my life; Ruby had been telling me that for a lot longer. She laid down the rules of life, she never minced words or dressed up a situation. Ruby had saved me from bad decisions many times over the years, so with her I would always listen.

Luckily Ruby liked Clare and told me to go for it. 'You're thirty-five, bald, and you've been sleeping on my couch for the last ten years . . . what, do you think you're Brad Pitt or something?'

With that she had a laugh—her laugh is priceless, like watching someone yawn. You find yourself laughing with her; throw in a few drinks and I'm on the floor, crocodile tears streaming down both cheeks.

'I'm going to get out of the oilfield soon,' I announced.

'Keep polishing that turd, Pauli.'

I spent more and more time with Clare. Her company was relaxing in a matter-of-fact way; I liked that. With nothing but time on my hands I occupied myself working for Louise in her advertising agency. I loved it, the people were fun and civilised, and the most danger I faced was a wayward paperclip or perhaps an overly hot cup of coffee.

After one particularly long stint offshore in the Philippines I returned home to a message on the answering machine, from Louise . . . Could I come straight over and have a superb dinner with lots of great wine and Barry stories. As usual it was a blast. The next day she sent me off to a photo shoot, my first. It was a studio shoot in town. I arrived, found the right studio and walked into a large white room filled with semi-naked girls, the photographer, the hair stylist, the make-up guy, who winked at me, and the client, who spent most of the time chain-smoking with his head out the window. Two days earlier I was on the drill floor on a shitty rig in the southern Philippines, and now I was getting paid to stand here and find inspiration.

In the three years that I have worked as a freelance copywriter in advertising, albeit in a random staggered way, I have had more drinks, dinners and parties thrown at me than in the fifteen years in the oilfield. But without fail after a few months I would get what my old boss called 'a rabbit up my arse' and I would start looking for a rig again. The characters you meet in the oilfield are unbelievable—from full-on rocket scientists with multiple Ivy League degrees and a keen interest in painting to-scale miniature sixteenth-century military figurines on their bunks, to Billy-Bob the brain-dead redneck ex-con whose misspelt jailhouse tatts, fart jokes and new truck back home are all he can talk about. Put a combination of twenty guys like that in a rundown backwater bar in some godforsaken corner of the world miles from anywhere remotely 'civilised', throw in a civil war, a donkey, and some festering prostitutes, and anything can happen. And I think that's why it's so addictive—not the drilling, not the job, definitely not the food, but the people and the situations you meet them in.

In the Philippines a few years ago I was in a bar with the boys when a gunfight started, yes, a gunfight. Everyone had a gun in that part of Manila. The time passed in super-slow mode, like recalling a car accident. But the part of that night that most sticks with me is when one of the guys went from drunk to sober in a second. We were hiding under the table together when I saw the first flash of panic on his face. Panic is a black leopard that sinks its claws deep into your skull; it makes your body burn and shake. Some people ball up, some freeze, some focus, I tend to poop my pants.

He focused, grabbed my collar and in a clear white moment said, 'If I get shot, you have to call my brother and tell him there's ten grand buried in a coffee can in his front lawn.'

I was blank . . . the guns were still going off, rounds were breaking windows and slapping into the wall somewhere above us. The fight lasted less than a minute, and spilled out into the street, where the two men exchanging fire were joined by two security men from the bar.

I took a quick look out the broken window; down in the street I could see one of the bar's armed bouncers crouched behind a car, his head almost in the wheel arch, one arm laying across the hood blindly firing in the general direction of the other men. I sat down on my unfinished cheeseburger and laughed a nervous relieved laugh. Mike was still under the table, lighting a cigarette. The others were scrambling to get out the back door.

'You buried ten grand in your brother's front lawn?'

'Fuck no, but he's a prick and it would have served him right.'

Towards the end of my fifth month at home I got a call to go back to Russia. I only had three days to get ready, and a bunch of copy to finish for Louise. It was a Sunday, my flight was on Monday morning, I was racing across town on the bike to get to a meeting. Just as I passed a set of traffic lights I sensed a car, way too close, and doing well over the speed limit. It came up behind me fast, hitting my bike's rear wheel as it was changing lanes.

The bike bucked, kicking me up on the tank. I held on but was overshooting a right-hand corner. I grabbed a handful of brakes but locked up the rear wheel . . . that was it . . . I laid it down. BANG . . . the impact was remarkably soft. I knew I was okay. I looked over my right shoulder as the bike and I slid down the road: the car that hit me was speeding off through the entrance to the Eastern Distributor Tunnel—bastard.

The corner . . . my head snapped back. The bike and I had parted company and I watched my beautiful

650 Twin slam into a metal fence. This can't be happening! I'd just had it resprayed.

Passing my demolished bike I bounced off a curb, somehow made the corner and slid through another set of traffic lights that were conveniently green. People sat behind their steering wheels watching me, mouths open, as I passed by them on my arse. I stopped just as my feet touched the curb of a cross-street. I got to my feet, but my right leg was shaking too much to stand.

My phone disintegrated in my hand when I opened it. A man came running up.

'Are you okay? Don't move, don't take your helmet off . . . your brains could fall out.'

I looked at the man and pondered that. 'Can I use your phone please?'

'I saw the whole thing, you're really fucking lucky mate.'

I called Ruby; she was on the way. An ambulance came roaring up, the medic sprang out. 'I'm okay . . . well, he thought the helmet was holding my brain in.' The medic shot a fuck-off look at the man who nodded in that knowing way.

'You'll live,' he said and smiled, then gave me some dressing for my right hand and soon they departed.

'Hope I never see you again,' I said.

Ruby arrived and took me to hospital, just to be sure. I'd cracked a rib.

'You can't go offshore now . . . shame, just gonna have to stay and be an ad man.'

Bollocks, I've never missed a job. The next day I was on the flight trying to work out how I was going to get into a survival suit, life jacket and four-point harness for the flight to the rig, let alone work.

(15)

AH MENG

2004

I WAS BACK IN Singapore again, begging for used thermal coveralls, but this time it was a different rig. BP had decided to drill that wildcat well offshore in Sakhalin. Erwin Herczeg was in charge of this job—the godfather. What a relief. No matter what blew up, broke down, fell over or just stopped on the rig, he could fix it, get the job done, do it in record time and have all the boys back on the beach in one piece and wearing the obligatory give-away client baseball caps. It had been a few years since we had worked on a job together; I was looking forward to it.

After ten days in the workshop standing by we finally got the call. This time we would go from Yuzhno by charter flight to a small jetty on the northernmost tip of the peninsula near the small town of Okha. From there we would take a supply vessel to a supply barge which was moored a few kilometres offshore in international waters. And finally a chopper to the rig.

By the time we arrived in Yuzhno the weather had turned. It was the middle of their summer but to us still cold enough to warrant thick jackets and beanies. The small charter aircraft was waiting on the tarmac, its strobes flashing and the door ajar. One prop was spinning, sending waves of invisible rollers over the grass behind us. We simply transferred from the jet to the charter. All the passengers were involved in the rig operation; one big Dutch guy who was with the company supplying the drill bits for the job was not happy with the look of the aircraft.

It did look as ancient as everything else on Sakhalin, like a flying version of the Nogliki train. The Dutch man walked around it kicking the tyres and swearing over the noise as the flight crew looked on. Eventually we all got on board. The bags were stowed by throwing them in the back; it was loud, cramped and very uncomfortable.

The weather had worsened by the time we arrived on the tiny Okha airstrip so the supply vessel was going to have to wait until the next day. Our accommodation for the night was a forty-minute truck ride inland, into the woods. It was very weird, we drove down a small dirt track ever deeper into an ominously black pine forest.

'Where the fuck are we going, there's nothing out here.' The only place I knew of for hundreds of miles was Okha, and we were heading away from there.

One of the BP guys leaned over. 'There are beds at this place, don't worry, it's not flash but it's all we could get at short notice.'

'What is it?'

'Oh, it's an old asylum.'

I looked at Erwin. 'Isn't that like a nuthouse?'

'Well it is now buddy,' he replied, laughing.

Darkness fell quickly, adding a sense of urgency that silently crept up everyone's spines. Then, towering above the woods, black against the sky, stood our lodgings. The building was another award-winning design from the Russian 'Fear Works' school of architecture. The only thing missing was a few well-placed gargoyles. It was a nuthouse. I pictured a Soviet version of *The Shining* with some crazed Boris hobbling about in the snow with a big fuck-off axe.

Random lights shone weakly in the upstairs windows, but when the truck circled and backed up to the main door, only its red tail lights illuminated the entrance. It started raining hard as we unloaded the bags, a dark figure opened the heavy iron door and right on cue a bolt of lightning cracked down over our heads. I wondered what this place must have been like during the Cold War, I thought about the possibility of salty-looking KGB film noir spies torturing people in the basement.

The woman in charge was perfectly suited to the spooky scene; Hollywood could not have cast this any better. She had a pronounced limp and actually said 'Walk this way' when she led us to our rooms on the second floor. I sniggered all the way up the stairs, along the barren corridors which were like huge dark tunnels.

I was sharing a room with Erwin; it smelled of disinfectant and looked like a cross between Hannibal Lecter's cell and your average public toilet. It was fairly empty, furnished only with two iron beds, each with one inch of antique foam, and a small table. I wished I had a crayon so I could write 'Red Rum' on the door. Every time lightning flashed through the curtainless window glass, thunder shaking the building, one of the guys down the hall would scream, 'IT'S ALIVE!'

Erwin sat on his bed, smiling. 'All work and no play makes Pauli a dull boy.'

'Very funny . . . I hope they feed us.'

Dinner was boiled mystery meat and boiled something else, yummy. We sat up late, swapping stories, catching up on the last few years while the wind beat a distinct rhythm through the rain on the cracked windows. We woke early to cold showers, clear skies and no breakfast. Then it was back on the truck to the jetty and the waiting supply ship.

Our voyage was brief, three hours and we were alongside the supply barge, our home until the rig was ready for us. The POB (Personnel On Board) situation on many rigs is a constant problem—no bed space. Sometimes I have had to 'hot bed' it, jumping straight into the night-shift guy's smelly sheets, too tired to care. To avoid this we would stand by on the supply barge until the last moment, then make the trip to the rig.

Inevitably there were problems with the drilling and we ended up standing by for a week. I was glad as it gave my rib a chance to heal. The vessel was comfortable, the people on board were really nice. Most of the crew were Indonesian, the rest Australian. But on the third day a massive typhoon started looming closer to our location. Tracking up the coast from Japan, it hit hard early in the morning. I knew this because I woke up on the floor with all my gear on top of me.

Remember those old *Star Trek* moments when the ship was being attacked and everyone would grab a handrail then collectively let go and do a high-speed dressage manoeuvre over to the opposite handrail while the director shook the camera? Well, a typhoon is nothing like that. If it's bad enough, it will pick you up and hurl you into the nearest wall. I could hear one of the boys throwing up with all his heart in the toilet.

'Don't open that fuckin' door.' I'm usually okay as long as I don't smell it.

Erwin was starting to turn green. 'I need to go on deck . . . see the horizon,' he moaned.

We pulled on raincoats and carefully made our way outside. The main deck was big, fifty square metres, and covered in equipment. Drums of drilling chemicals were scattered about and rolling all over the place. Containers had broken loose and skidded across the deck; the forklift lay on its roof. Every alternate wave broke over the side of the barge, crashing down hard on the cold steel floor. The hull vibrated as tonnes of sea water slammed like the wall of a liquid building into the bulkhead.

I tried to elevate Erwin's mind from that horrible inner focusing you do with seasickness. Waiting . . . swallowing . . . sweating on your alimentary canal to go into spasm. With one eye closed and my right leg tucked behind my coat, I hopped up to him. Horizontal rain stung my face, I had to yell as loud as I could over the wind. 'Arrrr me laddy, there we were, hard aground on the mahogany reef.' . . . Nothing . . . 'We was pickin' the weavels out the biscuits-n-drinkin' our urine all day, I tells ya.'

In the middle of my performance, a huge wave deposited a fully grown seal bang in the middle of the main deck. Equally surprised to see each other, all three of us exchanged looks and shared a 'What the fuck are you doing here?' moment. The seal remained firmly planted on the spot . . . ten feet from us, unexpectedly caught in the spotlight of the human world, his eyes wide as saucers darting left to right. I'd never seen a seal go from sheer terror to relaxed indifference . . . he almost smiled.

As if realising the most he had to fear was putting a

flipper in some wayward vomit, he opened his pink mouth and belched. Erwin and I watched him casually make his way over to the only shelter available, in the welding shop. Within an hour the boys were throwing him toasted sandwiches. I was waiting for him to come out and ask where we keep the good brandy.

The typhoon moved on, leaving days of cleaning up to be done. We made repairs and double-checked our equipment. No choppers were flying when our turn on the drill floor arrived, so we boarded the supply boat again. The rig, when we got there, loomed out of the fog, a decaying hulk, its structure forming an alien-fabricated atoll that reminds you of your fragility.

The crew stood on the deck of the supply boat, waiting in silence for the crane to lower the 'Billy Pugh'. Billy Pugh is a manufacturer's name, commonly used for a personnel basket. It looks like a giant upside-down ice-cream cone, with a flat ring about two metres wide and a cone of rope netting attached to the ring and fixed at the top to the crane's hook. Each man was running over the what-ifs in his mind. The first thing most of us usually do is take a good look at the derrick: Are there any stands racked back? Are they tripping in or out of the hole? What's my bunk going to be like? You just throw your bag inside the rope netting, step on to the ring with the other guys, grab the rope and hold on. The crane lifts you the 200-odd feet straight up and onto the heli-deck.

The concept is simple; up to four men at a time are hoisted to the rig. I felt sorry for one of the guys: he was pacing about, biting his nails, double-checking his survival suit, flashing occasional worried looks at the sky. If you do fall from the Billy Pugh you're stuffed; the water is so cold, survival suit or not, if the impact doesn't kill you the water temperature will. This guy was a young Russian wire line operator; I wandered over to him, smiling. He smiled back, rather like someone would with a gun to their head. Not only had the poor bastard been throwing up for the last nine hours in his cabin, he was now explaining in superfast broken English that he was scared of heights.

'You can ride with me if you like, you'll be fine,' I said in my best adult BBC serious English.

He nodded up and down quickly.

Our turn came, the bags were thrown inside the net, we stepped up, and grabbed on, the crane was taking up the slack fast as the boat was starting to heave.

'Look at me,' I said to the young Russian. He fixed his eyes level at mine. I smiled as we rose fast into the foggy airspace of salt spray and high anxiety.

Touching down he instantly lifted, resilient in his relief. I congratulated him.

'Not so bad,' he said, beaming.

The rig was a 600 Series semi-submersible; it had been sold just before this project kicked off so no one was interested in its shitty state. The whole thing was a mess;

from the top of the derrick to the Blow Out Preventer (BOP) deck it was a rusty eyesore that belonged in a graveyard, not hovering over a wildcat well.

Two weeks later we were all back on the supply barge, waiting for the next well section to come up. The barge, *Ismaya*, was built in Ireland in the 1950s and had gone through many roles in her long life. Converted into a drill ship in the early 1970s, *Ismaya* spent many years working in South-East Asia. Erwin had been on board when she was drilling in Indonesia.

One morning Erwin and I were having coffee when Garry the engineer joined us. He had worked with Erwin on the *Ismaya* so I learnt about some of the best years of her drilling past. According to them the rig's best crew member was 'Ah Meng', a young orphaned orangutan who the barge captain found in some harbourside market. He brought her back to the rig and there she stayed for many years.

One of the Indonesian crew members was a cabinet-maker. He built a bar room below decks, fashioned from beautiful teak, impossible to get now. When finished, it was the best rig bar in Asia; and it became Ah Meng's domain. She ran the bar on the rig for the next fifteen years. It was always clean and organised; she made cocktails. There was never any fighting because everyone had too much respect for Ah Meng . . . and if she wanted to she could pull your head off and throw it over the side. She had her stool; no one sat on her stool, ever. A favourite crew pastime was waiting for new people to come on board and unknowingly sit on Ah Meng's stool – only to be launched through the air into a large couch against the far wall. Even funnier if you could get a new guy to sit on the couch/landing area and another to plant himself on Ah Meng's stool.

Whenever the rig was in Singapore getting work done in dock, the boys would take Ah Meng out on the town. Erwin lived there for twenty years and remembered seeing her around now and again. Even the odd older taxi driver can recall having her in the cab. Since hearing the story I have asked every cab driver I met in Singapore and had three tell me she was just like any other tourist.

When the *Ismaya* was sold to a new drilling contractor, the company said Ah Meng had to go. The crew was in Singapore at the time, and decided to phone the Singapore Zoo for advice. The zoo had heard of Ah Meng, and immediately asked if they could have her. Apparently they said they would send a van over to the harbour, but were told that Ah Meng was on her way in a cab with the barge captain. The crew were in tears waving her off.

Barely in her twenties, she started a new life at the zoo. But Ah Meng was not to be paraded in an ordinary enclosure. Because of her bizarre circumstances and gentle nature, she became a kind of 'meet 'n greet' ambassador for the zoo. And as so many people have since told me, she is still there today and will probably continue to delight thousands of people for a long time. You can have your photo taken with her every day at lunch time. I made a date with her, hoping she would give me some tips on bar brawls and cocktails. She had a level gaze, summing me up in a second; I put my arm around her and wished she could talk.

EPILOGUE

NO MATTER WHAT HAPPENS in my lifetime and yours, we will always be involved in the oil business. Every time we start the car, heat the house, cook a meal, watch a war on the news, it reminds me that everything relies on fossil fuels to exist. Try not to think of the human cost, or the environmental cost. By 2080 we need a viable alternative to oil and gas, because by then one-third of our energy needs will have to come from somewhere else. Like solar power, wind power, geothermal power, hydrogen fuel cells, a genetically engineered three-storey hamster in a fuckin' huge wheel—I don't know.

In the meantime I'll keep drilling and writing bad copy. Who knows, I may even find normality . . . even marriage, children, a dog . . . who I will name Colin.

Get back to you in fifteen years.

ALSO BY PAUL CARTER

THIS IS NOT A DRILL
Just Another Glorious Day in the Oilfield

He's back on the rigs – and back in trouble!

"My work takes me to some strange places, usually Third World, and often during a coup, jihad, civil war, uprising, or riot of some description. If all that fails to happen then, with my track record, there will be a natural disaster. Only in the oil industry, the messy try-not-to-cut-a-limb-off side of the oil industry, does one realise first-hand that no matter what's going on in the world, the drilling goes on regardless – mind that landmine."

In this outrageous sequel to *Don't Tell Mum I Work on the Rigs*, Paul Carter picks up right where he left off, and pulls out more adventures from a mad, bad and dangerous life in the international oil trade.

Packed with action and mayhem galore, *This Is Not a Drill* cracks along at an unrelenting pace. In this fast, furious and very funny book, Paul almost drowns when the Russian rig he's working on begins to capsize; is reunited with his dad, another adrenaline junkie; gets married; hangs out with his rig buddies in exotic locations; gets hammered on vodka in Sakhalin; watches the winner of a crab race nip off his friend's toe; and spends a couple of interesting weeks in Afghanistan with some mates who run an outfit that just happens to contract out mercenaries for hire...

These are more great stories from the far side of civilisation, hilarious, full of humour and dramatic action!

UK £9.99 Paperback 978 1 85788 500 2
256pp 216x135mm
www.nicholasbrealey.com

IT'S ALL GREEK TO ME!
A Tale of a Mad Dog and an Englishman, Ruins, Retsina – and Real Greeks
John Mole

"Travel writing at its best. Mole's descriptions of the people and places he encounters do for Greece what Peter Mayle did for France in A Year in Provence and Frances Mayes for Italy in Under the Tuscan Sun."

www.greece.com

"A little whitewashed house with a blue door and blue shutters on an unspoilt island in a picturesque village next to the beach with a taverna round the corner…" This was the dream of the Mole family in search of a piece of Greek paradise.

But a beautiful view and a persuasive local prompted the impulsive purchase of a tumbledown ruin on a hillside with no water, no electricity, no roof, no floor, no doors, no windows and twenty years of goat dung…

In a charming saga of sun, sea, sand – and cement, John Mole tells of the back-breaking but joyous labours of fixing up his own Arcadia and introduces a warm, generous and garrulous cast of characters who helped (and occasionally hindered) his progress.

John Mole has had a varied international career, from banking in the USA and Athens to jacket potato restaurants in Russia. He is also a well-known author of comic novels and the perennial bestseller *Mind Your Manners*.

UK £7.99, US $14.00 Paperback 978 1 85788 375 6
352pp 198x129mm
www.nicholasbrealey.com

MISADVENTURES IN THE MIDDLE EAST
Travels as Tramp, Artist and Spy
Henry Hemming

"A once-in-a-lifetime journey, full of youthful ebullience and idealism, but self-aware too, and brave."
Colin Thubron, author of Shadow of the Silk Road *and* The Lost Heart of Asia

A beautifully written tale of a hapless artist, a truck called Yasmine and an extraordinary journey, creating a portrait of the post-9/11 Middle East that transports the reader into the human heart of the region and beyond the stereotypes.

Making art his passport, Henry Hemming's year-long travels take him from the drug-fuelled ski slopes of Iran via the region's souks, mosques, palaces, army barracks, secret beaches, police cells, nightclubs, torture chambers, brothels and artists' studios all the way to Baghdad and a Fourth of July party with GIs in one of Saddam's former palaces. From being accused of being an Islamic extremist to the Turkish army forcing him to make portraits of their girlfriends, from dancing in a dervish hideaway to being interrogated by the secret police as a British spy, *Misadventure in the Middle East* reveals an alternative Middle East that flies beneath the radar of the nightly news bulletins.

Henry Hemming is a British artist with a first-class degree in history from Newcastle University. He now works in a cramped East London studio and you can see examples of his work at www.henryhemming.com.

UK £10.99, US $19.95 Paperback 978 1 85788 395 4
304pp 214x135mm
www.nicholasbrealey.com

THUMBS UP AUSTRALIA
Hitchhiking the Outback
Tom Parry

"Tom Parry took on one of the marathons of hitchhiking – the round Australia route – and he has written a colourful and amusing account of the journey."
John McCarthy, Excess Baggage, Radio 4

Hitching lifts with the desert's dodgiest drivers and taking breaks in the roughest roadhouses, this is Tom Parry's witty, warts-and-all tale of hitchhiking 8,000 miles across – and around – the Australian outback with his thumb, his backpack and his French girlfriend Katia.

As the couple hitch their way around the near-empty highways they encounter as wide a cross-section of Aussie society as you could ever hope to meet. In cattle stations, Aboriginal communities, remote waterholes, caravan parks, hippy communes and roadhouses, they see a country that remains as extraordinary today as it was for the first nineteenth-century settlers.

Thumbs Up Australia features some of the country's most idiosyncratic characters, from the grizzled Aboriginal elder with his tales of dreamtime, to an amphetamine-swallowing roadtrain driver.

After trying to establish himself as a drummer in a rock band, **Tom Parry** is now a journalist with *The Daily Mirror* and has written travel articles for magazines such as *Traveller* and *Wanderlust*.

UK £9.99, US $16.00 Paperback 978 1 85788 390 9
288pp 214x135mm
www.nicholasbrealey.com